Ideas, Faiths,

Oxford University Press
Oxford London Glasgow
New York Toronto Melbourne Auckland
Delhi Bombay Calcutta Madras Karachi
Kuala Lumpur Singapore Hong Kong Tokyo
Nairobi Dar es Salaam Cape Town

and associate companies in
Beirut Berlin Ibadan Mexico City Nicosia

Library of Congress Cataloguing in Publication Data

May, Henry Farnham, 1915-
Ideas, faiths, and feelings.

1. United States—Intellectual life—Addresses,
essays, lectures. 2. United States—Religion—
Addresses, essays, lectures. I. Title.
E169.1.M4963 1983 973 82-18799
ISBN 0-19-503235-7
ISBN 0-19-503236-5 (pbk.)

Printing (last digit): 9 8 7 6 5 4 3 2 1

Printed in the United States of America

Ideas, Faiths, and Feelings

Essays on American Intellectual
and Religious History 1952-1982

HENRY F. MAY

New York   Oxford
OXFORD UNIVERSITY PRESS
1983

To My Students at Berkeley

Contents

Introduction

These essays have been selected from the short pieces I have written during the thirty years since I came to the University of California in 1952. Though I would write most of them differently now, I have left them almost as written. (Except for very minor stylistic changes, alterations are noted in the introductory statements that accompany each essay.) I have also presented them in the order in which they were written and published. I hope that in this way the series of essays may reflect the changing intellectual environment in which they were written, and the reaction to this environment of one historian.

The essays show the effect of several kinds of changes. Of these the most obvious though not I think the most important is the changing fashion among historians for different kinds of history. As a graduate student I worked under A. M. Schlesinger, Sr., who wrote and taught an eclectic mixture that he called "social and intellectual history." Though I greatly respected Schlesinger, I thought his kind of history neglected the analysis and interpretation of ideas. I wanted therefore to help develop a more independent kind of intellectual history, which would still not separate itself entirely from the history of society in the manner of A. O. Lovejoy and the history of ideas.

Quite a few other young historians shared approximately these purposes, and at about the time I came to Berkeley intellectual history was beginning a successful struggle for independence. It was under constant attack from older historians who thought it pretentious and insubstantial, and partly for this reason it attracted the allegiance of many graduate students. Intellectual history reached the height of its popularity in the 1950s and then in the sixties began to be attacked as "elitist" by spokesmen of various kinds of new social history. I found some of this attack valid, and some of the new social history highly interesting. From the middle of the sixties on I found myself increasingly interested in religious history, partly because it seemed a means of linking the history of ideas with the history of movements and institutions. I am not ashamed of my shifting loyalties among kinds of history, and indeed, have found these changes enriching.

In trying to remain flexible on these and other matters, I have been assisted by argument, vigorous and sometimes heated, with Berkeley graduate students. In these discussions, I continually changed my opinions about issues and methods, but gradually developed certain commitments to an idea of what history is. Whatever else it is, it is not an exact science. When I was a student, some working historians still assumed that a slow accumulation of facts would at some time automatically lead to unassailable truth. This belief, drawn from nineteenth-century natural science and carried over into the social sciences, has long been abandoned by historians, and for that matter by scientists as well. An approach to history drawn from pragmatic philosophy by historical thinkers of the early twentieth century works better. Historians can continually change their angle of vision and thus widen their picture. The picture gets richer and more complex but can never reproduce reality. In the long run, I believe, the historian is most like an artist, and history is irreducibly a kind of literature. However careful and ingenious and innovative his accumulation of facts, the historian must finally strive to present them in a way that engages the reader's emotional and aesthetic response as well as his intellectual assent. His art is as demanding and difficult as any other, and as in all the arts success is rare, elusive, and unpredictable.

The changing subject-matter of these essays has been affected by long struggles with books on different periods. My first book, published before the first of these essays was written, dealt with the late nineteenth century. The second, which I was working on through most of the 1950s, was an interpretation of the cultural history of the early twentieth century. From 1960 until 1976 I was trying to understand the late eighteenth century. It was my good fortune that my teaching always required me to range through all periods of American history. Some of these essays represent trial schemes of organization of particular periods, others treat themes running through several centuries and reflect my teaching. The first kind are usually addressed to historians, some of the others were written for unspecialized public audiences.

As I read through these pieces, I find that they reflect more than I had realized my changing views of the current scene, that is, my political views in the broadest sense. I have excluded directly political essays, but included two, essay 5 and review essay 2, that deal at once with historiographical and political concerns.

In the thirties I had been deeply affected by the populistic quasi-Marxism then in vogue. In the forties I was slowly discarding this influence, and doubtless this had something to do with my insistence on the autonomy of intellectual history. I always assumed that this autonomy was to be provisional and tactical, and never believed that people's ideas could be understood without reference to their lives and work. Perhaps this assumption had been lodged in my mind by early Marxist influences, perhaps by personal experience in and after the Depression.

In the fifties my writing clearly reflected the dominant climate of academic opinion, discussed briefly in essay 5. Absorbed in the problems of the expanding university, tolerating its tensions and enjoying many of its activities, I was also still reacting against early left-wing loyalties while trying to do my part in protesting McCarthyite excesses. Rejecting the simplicities of early progressive history, I probably came to place an unduly high value on complexity for its own sake. I certainly presented, both in teaching and writing, a largely positive view of American society. This is clearly reflected in several of these essays.

In the middle and late sixties my complacency was roughly

shaken by the student revolt. This event is discussed directly in essay 5 and its influence is reflected in the later essays. Without by any means identifying myself with this revolt, I was fascinated by its pervasive mood and its underlying assumptions, as powerful as they were elusive. I realized much better than I had earlier that history is determined not by ideas alone or by socio-economic conditions alone, but also by individual and group emotions that are seldom easy to explain fully.

In this period I learned how hard it is, in times of trouble, for the historian or anybody else to achieve the right balance between detachment and involvement. My efforts to achieve this balance, and thus my historical writing, were affected by one more personal change. From the late fifties through the sixties I found myself coming partway back, with much doubt and difficulty, to long-abandoned religious commitments. I tried, however, to keep my sympathy with religions past and present from biasing my work, and certainly retained my respect for some of the secular loyalties of the Enlightenment and its heirs.

It seems appropriate to dedicate these essays, written at Berkeley, to my students. I am deeply grateful to them for intellectual interchange and friendly discussion, developing in many cases into rich and sustaining friendship.

H. F. May

Berkeley
August 1982

Essays and Lectures

. I .

The Rebellion of the Intellectuals, 1912-1917

The first two of these essays were written during a long and only partly successful struggle to write a general book on the 1920s. This decade, the time of my childhood, was then only thirty years distant. Yet we were separated from it by the deep gulfs of Depression and World War. No historian had presented its cultural, political, and economic history in a single synthesis. Literary historians saw it one way; social and political historians another. It is always difficult to bring these two fields together, but especially so in dealing with a period when literary intellectuals were conscious of alienation from society.

In the process of trying to understand the literary and intellectual movements of the period I had been led to look closely at their origins in the years right before the war. With some excitement, I concluded that almost all the intellectual tendencies of the period had started before 1917. The war had changed them, but it had not caused them. Indeed, the profound impact of the war was partly caused by divisions that already existed. Concentration by historians of the prewar period on the dominant political progressivism had led them to ignore the quite different movements of intellectual revolt.

When I wrote this essay in 1954, I thought of it as a program for the first chapter of my book on the twenties. However, as I continued to work on the immediately prewar period I found it so complex and stimulating that it came to dominate my imagination. *The End of American Innocence* was published in 1959 with the subtitle "A Study of the First Years of Our Own Time 1912-1917."

It is perhaps instructive that neither the book nor this essay made any difference in most standard surveys, which continued to say that the war itself produced everything from intellectual revolt to jazz-age mores or fundamentalist reaction—all clearly visible before 1917.

This essay was read at the convention of the Mississippi Valley Historical Association in 1955 and published by the *American Quarterly*, VIII, 2 (Summer 1956), pp. 114-26. It is reprinted by permission of the editors and is copyright, 1956, by the Trustees of the University of Pennsylvania.

As the nineteen-twenties move from memory into history, a standard picture of the decade emerges from reminiscence and research into the textbooks. This picture is a puzzling one. The decade is still, as it was in the thirties, the last island of normalcy and isolation between wars and crises. Yet it is also, clearly, a period of major cultural revolution. Both the "revolt of the highbrows" and the "rebellion of youth," first sketched by Frederick Lewis Allen, are a standard part of our semiofficial self-picture. In response to current historical fashions and perhaps also to their own changing worries about their country, historians are giving more attention to the revolutionary aspect of this conservative decade.

Having dealt with other revolutions, historians should be able to appreciate both the importance and complexity of this one. For instance, they should be able to avoid taking to task the rebellious intellectuals of the twenties in the manner of some critics of the forties. The spokesmen of a revolution are not, after all, its sole cause, and a healthy regime is seldom overthrown. Yet anybody, today, must recognize that revolutions are expensive. They may release, as this one certainly did, a burst of creative vigor; but they inevitably leave behind division, hatred, and shock. In the twenties, for instance, beliefs and customs that still commanded the deepest loyalties of one part of the population became to another group a dead and repressive Genteel Tradition, to be ceremonially flouted whenever possible. Suspicions dating from this cultural cleavage still poison the air. The historian must hope that analysis of the revolution and its causes can eventually help a little to clear away some of the resentment.

Starting backward, as historians must, we arrive immediately at the First World War, and there many have stopped. It is obvious that America's first major venture into world tragedy, with its rapid cycle of national exaltation, exhaustion, and revulsion played a very large part in the emotional life of the postwar rebels. By contrast with 1918 or 1919 or even 1925, hundreds of autobiographies paint the prewar period as a time of unity, moderation, progress, and sheltered childhood.

Yet we all know that postwar reminiscence, whether of the old plantation or the old literary culture, is a dubious guide for history. Those who have looked even briefly at the social and literary criticism of the prewar years know that the period 1912-17[1] was itself, for some, a time of doubt and fragmentation, of upheaval in ideas, of the disintegration of tradition—in other words it was a pre-revolutionary or early revolutionary period. Nearly every phenomenon of the twenties from Freudianism to expatriation or the abandonment of politics was present before the war, on a smaller scale and with certain differences. If we can recapture any of the meaning or content of this prewar ferment, we may be able to understand better in what sense the revolution of the twenties was and was not postwar. In this way we may even get a few suggestions as to the perenially baffling problem of the relation between ideas and events.

In an essay published in 1913 George Santayana made an attempt to catch and pin down on paper "The Intellectual Temper of the Age." To do this for one's own time is one of the hardest tasks a writer can undertake, yet for anybody who has been for a while immersed in the records of that period it is astonishing how well this brilliant essay seems to order and illuminate the times. To Santayana it seemed that "the civilisation characteristic of Christendom has not disappeared, yet another civilisation has begun to take its place."[2] In the resulting age of confusion and transition, men were giving up the search for lasting values and firm intellectual conclusions. Instead of these, they were placing a premium on sheer vitality, on movement, change, and emotion. According to Santayana, who sometimes enjoyed but did not admire this taste, the result was that in thought and art, his generation was "in full career toward disintegration."[3]

Whether or not one shares Santayana's cool disapproval of the tendencies of his day, the vitalist spirit he describes stares out from the sources. One recognizes on all sides its gaiety, its irresponsibility, its love of change, and also its contempt for reason. And it does not take much knowledge of American intellectual history to know that this spirit meant trouble. For a century and a half the dominant ideas in the national faith had been a confidence in secure moral values and a belief in progress. These two commitments had often been in conflict and formed at best a somewhat unstable compound. Now both at once were brought under devastating attack.

If one starts, as Santayana does, with philosophy, the tendencies he describes emerge very clearly. The young intellectuals of America were still most widely influenced by pragmatism, by what Morton G. White has called the revolt against formalism.

Experience and movement were reality; potentiality more important than actuality. Dewey's program for intelligence remaking both the world and itself probably attracted the largest number of conscious disciples, some of them, like Randolph Bourne, soon to break away in a more emotionally satisfying direction. But it may well be that the influence of James, with his catholic and dangerous acceptance of the irrational, personal, and mysterious went deeper in a generation nourished on idealism. Emerson, universally read though misunderstood and underrated, and Whitman, the sole American patron of some of the rebels, as well as the German idealists casually studied in college courses, must have prepared them for a philosophy of intuition. Whatever the reason, it was the destructive elements in pragmatism that were the most influential. The avant-garde of 1912-17, the aggressive young innovators, were perfectly willing to see all of life as an experiment. But their purpose in experimenting was rather to express themselves and experience emotion than to solve problems in a disciplined manner.

Those who were sensitive to Atlantic breezes felt most keenly the swelling winds of antirationalism, which had been gathering force for a long time. Nietzsche, for long known vaguely by the American public as an Anti-christ, was becoming a major prophet. The most vigorous, though not the most accurate, of his American interpreters was H. L. Mencken, who in a widely read and widely praised book published first in 1908 and again in 1913 used the German prophet to belabor religion, women, and, most roughly of all, democracy in his already familiar manner.[4] But the most fashionable of Europeans was the still living and lecturing Henri Bergson, who pushed the current tendency to an extreme, contending that reality, being in constant flux and change, is only distorted by efforts to think about it and must be apprehended through intuition. His was not the only, but it was probably the dominant direction in which philosophy was moving in 1913, and there is plenty of evidence that he was extraordi-

narily attractive to up-to-date American intellectuals. Irving Babbitt, already an alarmed defender of traditional values, saw the
rise of Bergsonism as the culmination of a long, deplorable irrationalist trend, and found it in 1912 "allied with all that is violent and extreme in contemporary life from syndicalism to 'fu-
turist' painting."[5]

Psychology, as well as philosophy, was dealing heavy blows to
dominant assumptions and beliefs. From the time of Freud's famous trip to Clark University in 1908, the Viennese theories
cropped up in popular magazines and political treatises as well
as learned journals. Whether or not, as his supporters claim,
Freud is to be regarded as himself a brave and determined cham-
pion of reason, the first impact of his doctrines in the United
States seemed to confirm and deepen the hedonism, emotional-
ism, and egocentricity that were beginning to spread so widely.[6]
On the other hand, Behaviorism, a movement launched in its
most dogmatic form by John B. Watson in 1912, had to wait for
its vogue until after the war.[7] Its extreme practicalism, its rejection not only of reason but of consciousness, its suspicion of emotion, did not fit the tastes of the prewar rebels.

It does not need demonstrating that restless and vigorous innovation in the graphic arts got its American start before the war.
Two major tendencies already dazzled the intellectuals and star-
tled the public. One was apparently native, the harsh and sometimes violent Ash Can realism of Sloan, Bellows, and the *Masses*
cartoons. The other was imported from Paris and consisted of a
kaleidoscopic series of schools of experiment in form and technique. Commenting on "Current Impressionism," a term already
well out of date but helpful as a catch-all, Louis Weinberg extended his observations from and beyond contemporary art:

> Impressionism as a technique is a means of recording the transi-
> tory nature of phenomena and the fluidity of motion. As a
> principle it is based on a philosophy of change. . . .
> But this is not alone a description of the art of our times. It is
> the very essence of our lives.[8]

Wherever the impressionist or vitalist tendency arose, it was expressed most frequently and characteristically not in painting or philosophy, but in politics and literature. These are the forms in which most American cultural change has been recorded, and it is to them that we must turn for a slightly closer look at prewar tendencies. Santayana's brilliant summary suggests that in politics alone the current drift toward fragmentation and chaos may have reversed itself in the constructive and integrating (though to Santayana most uncongenial) movement towards collectivism.[9] In this one opinion, regarding an area which concerned him little, I think Santayana missed the current drift and underrated the force of his own generalization. It is true that progressivism, optimistic, gradual, and in some forms mildly collectivist, was the officially dominant ideology; and that socialism was a swelling movement on the left that seemed to many sober Americans to possess the future. Yet both these political tendencies were in the early teens already under devastating attack, and from much the same irrationalist quarter.

Progressivism in all its varieties took for granted one or both of the two fundamental assumptions which had so far underlain the whole American political tradition. One of these was that we possess secure criteria by which we can judge our political achievement, the other that human beings are able consciously to remold their environment. Now both of these basic assumptions were being seriously shaken by new doctrines that had penetrated the house of progressivism itself.

Recent studies have shown that moral standards of a highly traditional sort motivated a great many of the prewar progressives. Truth and falsehood, good and evil, stand out in nearly all the speeches of Theodore Roosevelt and Wilson, and good men threw out bad in most American cities. These venerable distinctions were the first to go; the younger progressive intellectuals, nourished on Dewey and H. G. Wells, were quite willing to throw out moral categories and rely on the shaping intelligence. On a popular level Lincoln Steffens spread the picture of the good boss and the honest crook. James Harvey Robinson, speaking for the main organ of the pragmatic progressives, lumped together as obsolete the ideals of "sound doctrine, consistency, fidelity to conscience, eternal verities, immutable human nature, and the imprescriptable rights of man."[10]

With these went the state and law, the traditional locus and method of American reform. Many of the ablest political theorists of various schools, led by the brilliant Harold Laski, were redefining the state almost out of existence. To some it was a congeries of associations, to others the tool of a class, to still others the expression of the wish of those at present in authority. Its acts were no more final and deserved no greater obedience than those of other human groups, and it was less likely than many to be rationally motivated. Similarly, law, to the followers of the French positivist Leon Duguit or the American Roscoe Pound was no longer either the embodiment of a principle nor the command of a sovereign, but the complex resultant of social forces, prevailing opinion, and judicial will.

There remained the conscious intelligence, remolding the goals of action together with its methods. This was a moving conception, and a sufficient loyalty for many in this generation. Yet this too was seriously menaced by ideas that were attractive to the youngest generation of progressives. From the new and flourishing disciplines of sociology, anthropology, and social psychology came an increasingly fashionable emphasis on custom and group emotion. It was sometimes hard to see what function this newest tendency left for intelligence and purpose.[11]

Walter Lippmann's two prewar studies, *A Preface to Politics* (1913) and *Drift and Mastery* (1914) bring together the pragmatist attack on tradition and the implicit Freudian attack on pragmatism. Appealing for a radically instrumental state, he denounces the "routineers" who rely on political machinery, law, and conventional morality. His fellow progressives seem to draw most of his fire for their naïve adherence to literal numerical democracy and narrow utilitarian goals. What is needed in politics is passion and creative emotion, still of course somehow constructively channeled and used by the far-seeing for purposes which will transcend woman suffrage or the eight-hour day.

> . . . the goal of action is in its final analysis aesthetic and not moral—a quality of feeling instead of conformity to rule.[12]

This formulation seems to me far closer to the view of postwar literary intellectuals than to that of the progressive standard-bearers. And the sources are explicit. Lippmann's friend Graham Wallas, the British author of *Human Nature in Politics*[13] had opened the eyes of his Harvard seminar to political psychology. Steffens had helped to guide Lippmann and so, in a negative direction, had his brief experience with municipal socialism in Schenectady. But beyond these immediate guides one finds recurring references to James, Nietzsche, and Bergson and frequent, specific acknowledgment of the work of Freud.[14]

All these new insights enriched the social sciences, and for many they doubtless furnished in practice new sources of power and freedom. Traditional progressivism, with its facile assumptions and sometimes shallow purposes needed—and for that matter still needs—rethinking. Yet much had been accomplished under the auspices of ideas that were beginning to seem stale and boring. And the new beliefs that buzzed and swarmed through the immediate postwar years were not easy to introduce into the progressive hive. To combine Lippmann or Laski with Wilson was, and soon proved to be, as difficult as to match Bergson and Nietzsche with Lyman Abbott.

It is tempting to wonder whether the actual practical difficulties of progressivism from about 1914 to 1917 were not related in part to confusion of purposes and motives. It is true at least that the Wilsonian impetus began to bog down in these years. Already one finds in the up-to-the-minute *New Republic* troubled editorials that ask the common postwar question: what has happened to the progressives?[15]

On the far left much the same process was taking place, whether one labels it fertilization or disintegration or both. Not the Marxian dialectic, but the Bergsonian and mystical syndicalism of Sorel or the anarchism of Max Stirner or Emma Goldman seemed most exciting to the younger radical intellectuals.[16] Not the earnest socialism of Milwaukee or Schenectady, with its respectability and its reliance on the discredited machinery of the state, but the romantic activism of the I.W.W. captured the emotions of the sympathizers. One of America's waves of labor violence, running through the Northwest, Colorado, West Virginia, and other centers of direct action, reflecting the primitive brutality of employers' methods in the same areas, aroused the generous emotions and seemed to some to make political action irrelevant. The climax came in 1912 at Lawrence and in 1913 at Paterson, when the I.W.W. penetrated the East and the writers and artists went to its aid, when Bill Haywood was a Greenwich Village social lion and John Reed staged an immense pageant in Madison Square Garden with the letters I.W.W. flaming from the roof in red electric signs ten feet high. Even Lippmann, viewing radicalism from the outside, approved the I.W.W. rather than the Socialist Party as less formalist and more in possession of the kind of emotional force that needed only to be constructively channeled.[17]

Naturally, when Max Eastman, a young man of impeccable ministerial stock, joined the Socialist Party, he chose the left wing rather than the gradualists. Under Eastman's editorship the *Masses,* focus of so much later radical nostalgia, became perhaps even more clearly than the sober *New Republic* the organ of youth. Publishing the magnificent and not always political cartoons of Sloan and Bellows, an occasional Picasso drawing, stories by Sherwood Anderson, and reporting by Reed, it fought for the new literature and the new sexual morality as well as the social revolution. The *Masses* was rich in humor and human emotion— qualities often enough lacking on the far left—and practically negligible in social program. Smashing idols was, in these years as after the war, a flourishing business, while Socialism as a political movement was already losing momentum in 1914-16.[18]

More spectacularly than anywhere else, the new spirit of 1910 or 1912 to 1917 was reflected in a literary renaissance. The story of this sudden creative outburst has often been told, and only two points need making for our present purpose. One of these is that literary departures in the prewar years were closely related to contemporary movements in other fields of thought, the other that prewar writing contains in embryo nearly all the developments of the twenties.

Here too the stimulus came in large part from abroad. Young Americans, brought up on Matthew Arnold and Thackeray, were following before he gave it the advice of Yeats at the *Poetry* dinner in 1912 to forget London and look to Paris for all that was excellent.[19] In Kroch's bookstore in Chicago, in the translations issued by a series of daring new publishers, in the eager if undiscriminating reviews by the young critics, this generation of rebels was nourished on a whole series of movements extending over the last generation in Europe. All the writers that had for so long been belaboring the European bourgeoisie—French symbolists and decadents and naturalists, Scandinavian pessimists and social critics, Russian apostles of mysticism and emotion; even from England D. H. Lawrence as well as Shaw, suddenly began to penetrate the American barrier. What this series of reagents produced was a series of explosions, and what exploded was more than the genteel tradition in literature, more than conventional moral categories. With the conventions of literary form and language went the underlying assumptions about thought and communication. Randolph Bourne perhaps described this grand explosion better than he realized in June, 1917:

> What becomes more and more apparent to the readers of Dostoevsky, however, is his superb modern healthiness. He is healthy because he has no sense of any dividing line between the normal and the abnormal, or even between the sane and the insane.[20]

When Harriet Monroe, full of civic feeling as well as poetic zeal, founded *Poetry* in 1912 she seemed to tap immediately a rich underground gusher of poetic impulse. Soon the flood of experiment became too thick and varied even for *Poetry* to contain and overflowed into *Others* and the *Little Review*. As in the visual arts, a rapid series of schools succeeded each other, but perhaps the literary movement most characteristic of the period, and most obviously related to its philosophic tendencies was that of the Imagists, with its manifestoes in favor of complete freedom, concentration on the fleeting and immediate image for its own sake, and refusal to assign an image any "external" meaning or reference. Already before the war the revolution in the use of language was under way toward its ultimate destinations; Joyce was being published in the London *Egoist* and Gertrude Stein, settled in Paris, had developed her opinions and her characteristic style.

It would be misleading to divide this literary outpouring into precise categories, yet one can find suggestions of two emergent ways of thinking and feeling among writers. One group demanded freedom from European forms, confidence in emotion and spontaneity, and in general preached democratic optimism in the Whitman tradition. The other, more disciplined but also more deeply rebellious against American culture, called for concentration, rejection of irrelevant moral and political purposes, and the development of conscious intellectual aristocracy.

Obviously the former, democratic and optimist group is more distant than the other from postwar directions. This is the tendency one associates particularly with the sudden and brief Chicago Renaissance, with Sandburg and Lindsay and Miss Monroe, though it is found also in other areas, for instance in the organized and vigorous character of what Francis Hackett labeled and dated forever as Miss Amy Lowell's "Votes for Poetry movement."[21] Yet even the most exuberant of the Chicago poets were, like contemporary political radicals, destroying for the sake of redemption, like Sandburg's personified city "Shovelling, wrecking, planning, building, breaking, rebuilding."

And even in Chicago pessimistic and sceptical tendencies were also, and had long been, at work. Dreiser's not exactly rosy picture of American city life was finally finding its audience; and the small town, from E. A. Robinson's New England Tilbury town to Masters' Middlewestern Spoon River, was preparing the way for Winesburg and Gopher Prairie. In the bosom of *Poetry* magazine, at the official heart of the Chicago movement, Ezra Pound, the magazine's foreign editor, was chafing at its cover slogan, the statement of Whitman that "to have great poets there must be great audiences too." Pound preferred Dante's pronouncement that the wisest in the city is "He whom the fools hate worst" and denied that poets have any need for the rabble.

> It is true that the great artist has always a great audience, even in his lifetime; but it is not the *vulgo* but the spirits of irony and of destiny and of humor, sitting with him.[22]

In that sentence lies the germ of a dozen ponderous manifestoes of the postwar Young Intellectuals. Pound stayed on *Poetry* long enough to persuade Miss Monroe to publish Eliot's "Prufrock" in 1915 and then found a refuge from uplift and Whitmanism in the *Little Review*.

In the Eastern centers of the new literary movement the mixture of optimism and nihilism, of reform and rejection was somewhat different. Harvard, which was incubating an extraordinary number of important writers, seemed to produce a strange and special mixture of ideas.[23] The dominant note in its teaching of literature was aestheticism, worship of Europe, and contempt for the native production. Irving Babbitt's vigorous attack on democratic sentimentality was already a major influence. Yet Walter Lippmann, for one, managed to combine presidency of the Harvard Socialist Club with assisting Professor Santayana. A certain survival of Emersonian and Puritan responsibility seems to have been a part of the prevalent passionate concern for literature. America might be vulgar and materialistic and nearly hopeless; if so one's duty was to search the harder for seeds of literary springtime, and literary revival would bring social regeneration as well. Like so many writers after the war, Van Wyck Brooks went to Europe to look for these seeds. He found in London in 1913-14 Ezra Pound, T. S. Eliot, John Gould Fletcher, Conrad Aiken, Elinor Wylie, Robert Frost, and Walter Lippmann.[24] Across the Channel he could already have run into an equally significant group of fellow-countrymen. It was in London that Brooks began to struggle seriously with the typical problem of the expatriate of the next decade: the love of European tradition and the nostalgic turning toward American vitality. He solved this problem by writing, in London in 1914, the book that most influenced the writers of the next decade, an attack on the Genteel Tradition and an appeal for a literary renaissance that seemed then, as its title implies, to mark an arrival and not just a beginning: *America's Coming-of-Age.*

From here we can see, even more clearly than Santayana could in 1913, the unrest, the disintegration of old standards, the search for vitality and movement that already was under way at that time.[25] We know, too, that what was then unrest later became cultural revolution and angry intellectual civil war. This brings us to the compelling question, what started it all? Why did this search for novelty, this gay destruction of traditional standards, occur at just this moment in the midst of an apparently placid and contented period?

This is hardly a question that can be answered with certainty. All that we know for sure is that a movement so general and noticeable in the prewar years was not started by the war. Perhaps the most obvious forces at work in early twentieth-century civilization were technological change and urban growth, but these had been at work reshaping American culture for several generations and do not afford a direct and simple explanation for the sudden restlessness of 1912-17. Moreover, an increase of mechanistic materialism rather than a new vitalism would seem a more easily understandable product of machine civilization. It may be that the prewar rebellion was in part a protest against such a long-run tendency; in 1915 the *Nation* suggested that the rising "Bergsonian school . . . owes not a little of its popularity to its expression of revolt from the dreary materialistic determinism of the closing years of the last century."[26]

One is tempted to wonder whether the new physics was at work already disintegrating the comparatively simple universe of nineteenth-century science. It seems, however, that although the Einstein revolution was being discussed before the war by American scientists and reported in the serious periodical press, it did not directly affect as yet the literary and political intellectuals to any great extent, and it was not, as it became after the war, a newspaper sensation.[27]

In part the American intellectual rebellion may be considered merely a belated phase of an European antirationalist tendency. Yet it remains puzzling that Nietzsche and Dostoevsky and Baudelaire waited for their most telling American impact until they competed with Freud and Joyce. Part of the violence of the American literary and intellectual battles of the next decade arises from the fact that influences that had gradually permeated European thought presented themselves to many Americans all at once and in their extreme forms.

The time and special character of the prewar rebellion were, I believe, determined in part by the very surface placidity of the Progressive Era. Traditional American beliefs in moral certainty and inevitable progress had for some time been subjected to inner strain and external attack, yet in the prewar decade, for the last time, the official custodians of culture were able to maintain and defend a common front. Yet these almost hereditary leaders—Roosevelt and Royce and Howells in their several spheres—were growing weaker. A new generation, full of confidence and provided with leisure and libraries, was fairly literally looking for trouble. What attracts us about the standard culture of America in the early years of the century is its confident consensus, its lack of passion and violence. Passion and violence were exactly the demand of the young intellectuals of 1913 and 1914, of Lippmann and Brooks and Bourne and Pound. This was what they wanted, and this was what they got.

The war, then, was not the cause of the cultural revolution of the twenties. It played, however, the immense part that the Civil War played in the economic and political revolution of the mid-nineteenth century, speeding, widening, and altering in kind a movement already under way.

The experiences of 1917-19 darkened and soured the mood of the rebels. Even at its most iconoclastic and even in those spokesmen who adopted the most pessimistic doctrines, the prewar renaissance was always exuberant. Pound, amid his fierce negations, still found it possible to make his famous and somewhat rash prophecy that the coming American Risorgimento would "make the Italian Renaissance look like a tempest in a teapot!"[28] The rejection of easy rationalism, the spurning of dull politics were to make America better and brighter. In the war and its aftermath however the rebellious generation learned something of the price of destruction and experienced personally both tragedy and (in 1919) failure. Many who had been rebellious optimists became despairing nihilists and those who had already been preaching the futility of conscious effort preached it with different emotional corollaries.

The other effect of the war was that the disintegration of traditional ideas spread far more widely among the population. Most of the prewar rebellion was confined to a small and isolated, though articulate and potentially influential, group of intellectuals. As yet the overwhelming bulk of the people took for granted the truth of the old political and moral slogans. As long as this was so rebels could be ignored or patronized; they did not have to be feared and fought. Without the political events of 1917-19 traditional beliefs might perhaps have been slowly adapted to new realities instead of, for the moment, either smashed or blindly defended. And without the currents of doubt and disintegration already abroad, these political events themselves might have lacked their willing and ready Cassandras.

In 1913 *Sons and Lovers, A Preface to Politics,* and *Winds of Doctrine* were published, but *Pollyanna* and *Laddie* were the best-sellers. In 1925 the best-seller list itself had to find place for *An American Tragedy.*

Notes

1. Through this essay I treat this period as one instead of dividing it in August, 1914. The outbreak of the war in Europe shocked American intellectuals but did not immediately become their main preoccupation. Until about the winter of 1916, radical and progressive politics, together with the new literary and philosophical tendencies, get more space than the war in the liberal and literary periodicals.
2. George Santayana, *Winds of Doctrine* (London and New York: Charles Scribner's Sons, 1913), p. 1.
3. *Ibid.,* p. 10.
4. Henry L. Mencken, *The Philosophy of Friedrich Nietzsche* (Boston: Luce and Co., 1918).
5. Irving Babbitt, "Bergson and Rousseau," *Nation,* November 14, 1912, p. 455. One of the more influential of the considerable number of books on Bergson appearing in these years was H. M. Kallen, *William James and Henri Bergson* (Chicago: University of Chicago press, 1914). There is a very large volume of periodical discussion from 1911.
6. For a helpful review see Frederick J. Hoffman, *Freudianism and the Literary Mind* (Baton Rouge, La.: Louisiana State University Press, 1945). The early impact of Freud and many other foreign influences is clearly recorded in the works of Floyd Dell, one of Freud's important American exponents. Dell deals most specifically with these influences in his retrospective *Intellectual Vagabondage* (New York: George H. Doran Co., 1926).
7. See Lucille C. Birnbaum, "Behaviorism in the 1920's," *American Quarterly,* VII (1955), 15-30, esp. p. 20.
8. Louis Weinberg, "Current Impressionism," *New Republic,* March 6, 1915, pp. 124-25.

9. George Santayana, *Winds of Doctrine*, p. 10.

10. James H. Robinson, "A Journal of Opinion," *New Republic*, May 8, 1915, pp. 9-10.

11. An account of all these tendencies in prewar thought, together with a vast bibliography, can be found in two helpful summaries. These are W. Y. Elliott, *The Pragmatic Revolt in Politics* (New York: The Macmillan Company, 1928) and C. E. Merriam and H. E. Barnes, eds., *A History of Political Theories, Recent Times* (New York: The Macmillan Company, 1924).

12. Walter Lippmann, *A Preface to Politics* (New York: M. Kennerley, 1913), p. 200.

13. London: A. Constable and Co., 1908.

14. *E.g.*, Walter Lippmann, *Drift and Mastery* (New York: M. Kennerley, 1914), pp. 249, 274.

15. *E.g.*, January 16, 1915, pp. 6-8; November 6, 1915, p. 1; June 17, 1916, pp. 159-61; July 1, 1916, pp. 211-13.

16. See Daniel Bell, "Marxian Socialism in the United States," in D. D. Egbert and Stow Persons, eds., *Socialism and American Life* (Princeton, N.J.: Princeton University Press, 1952), I, 289-90.

17. Walter Lippmann, *Preface to Politics*, pp. 277-78.

18. See David L. Shannon, *The Socialist Party of America: A History* (New York: The Macmillan Company, 1955). As Shannon and other historians of socialism have pointed out, the apparent revival of the Socialist Party in the big Debs vote of 1920 is misleading. It belongs in the category of protest rather than party success.

19. Harriet Monroe, *A Poet's Life* (New York: The Macmillan Company, 1938), p. 337.

20. Randolph Bourne, "The Immanence of Dostoevsky," *The Dial*, LXIII (1917), 25.

21. In the *New Republic*, November 10, 1917, p. 52.

22. Ezra Pound, "The Audience," *Poetry, A Magazine of Verse*, V (1914-15), 30.

23. See the following helpful autobiographies of Harvard graduates: Malcolm Cowley, *Exile's Return* (New York: W. W. Norton & Company, 1934); Harold E. Stearns, *The Street I Know* (New York: L. Furman, 1935); Van Wyck Brooks, *Scenes and Portraits* (New York: E. P. Dutton & Co., 1954).

24. Brooks, *op. cit.*, pp. 123-48, 210 ff.

25. The same traits that one finds in the ideas of the period characterize much of its social life. Ragtime and the dance craze, the

furor over alleged white slave disclosures, in 1913 seem to prefigure the feverishness and the moral divisions of the postwar decade.

26. From a review of Croly's *Progressive Democracy*, which the *Nation* associates with the Bergson influence (April 29, 1915), pp. 469-70.

27. This impression comes from an examination of periodicals and is confirmed by an intensive though brief examination of popular scientific literature by Robert G. Sumpter.

28. Pound to Harriet Monroe, 24 September, 1914, in D. D. Paige, ed., *The Letters of Ezra Pound 1907-1941* (New York: Harcourt, Brace and Co., 1950), p. 10.

· 2 ·

Shifting Perspectives on the 1920s

Of the essays in this book, this is the one I now like least. Its faults can I think be explained in several related ways. First, it is a specimen of the familiar tactic of avoiding historical difficulties by sticking to historiography, dealing with what historians say rather than coming to grips with what happened. Second, it was presented before an organization with very mixed feelings about intellectual history. Thus its tone is in places that of an apologist for that field, trying to appease all possible critics. Finally, it is a prime example of the historical tendencies of the 1950s—the social and economic optimism, the admiration of complexity for its own sake, and especially the assumption that we understand history best by avoiding passion.

I chose to include this essay in the present collection because it does make a strenuous effort to bring together a number of approaches to a puzzling decade. No historian has yet succeeded in writing a study of the twenties as comprehensive as that suggested here. Perhaps this essay will at least help to show why.

Probably because this essay is so much more conventional than its predecessor, it proved much easier to publish. It was prepared for the 1955 meeting of the American Historical Association and published in the *Mississippi Valley Historical Review*, XLIII, 3 (December 1956), pp. 405-27. It is reprinted by permission of the editor.

To comment on the 1920s today is to put oneself in the position of a Civil War historian writing in the 1890s. The period is over and major changes have taken place. The younger historian himself belongs to a generation which barely remembers the great days. From the point of view of the veterans, still full of heroic memories, such a historian obviously has no right to talk—he was not there. Yet historians are led by their training to hope that one kind of truth—not the only kind and perhaps not always the most important kind—emerges from the calm study of the records.

Calm study of this decade is not easy. Like the Civil War itself, the cultural battles of the twenties have been fought again and again. Successive writers have found it necessary either to condemn or to praise the decade, though what they have seen in it to condemn or praise has differed. Perhaps this fact offers us our best starting point. If we can trace the shifting and changing picture of the decade through the last thirty years, and still better, if we can understand the emotions that have attached themselves to one version or another, we may be closer to knowing what the decade really meant. In the process, we can hardly help learning something of the intellectual history of the intervening period.

It is immediately apparent, as one turns through the literature about the twenties, that most of the striking contributions have not come from men we usually think of as historians, but rather from journalists, literary critics, and social scientists. This is perhaps not surprising, since most of the excitement has centered in

areas outside the historian's traditional domain. Historians today, of course, claim a territory stretching far beyond past politics; but this is a recent expansion, and all of us enter such fields
as literature and science only with caution. Caution is necessary,
but it must not prevent exploration. If the best insights into a
period come from economists, or anthropologists, or literary
critics, we must try to understand and even to assess them, hoping
that our inevitable mistakes will be made in a good cause.

At least three pictures of the twenties had formed before the
decade was over. For different reasons, spokesmen of business,
social science, and literary revolt all wanted to get clear away from
the past, to discard history. For this reason, all three groups were
constantly discussing their own historical role. Perhaps the dominant current version was that proclaimed by the businessmen, the
picture of the period usually conveyed by the phrase New Era itself. Out of the postwar upheaval was emerging, in this view, a
new civilization. Its origin was technology, its efficient causes high
wages and diffusion of ownership, its leadership enlightened private management. This picture of the period was far more than a
matter of political speeches and *Saturday Evening Post* editorials.
It was buttressed by academic argument and attested by foreign
observers. To its believers, we must remember, it was not a picture of conservatism but of innovation, even, as Thomas Nixon
Carver strikingly asserted, of revolution.[1]

It is not surprising that this interpretation of the period gained
the allegiance of many of its first historians. Preston W. Slosson,
surveying his own time for the *History of American Life* series,
came to a typical New Era conclusion on the basis of a typical
New Era criterion: "Often in history the acid test of wealth has
been applied to a favored class; alone in all nations and all ages
the United States of the 1920s was beginning to apply that test
to a whole people."[2] James C. Malin found, with no apparent
anguish, that political democracy was being replaced by self-
government in industry.[3] No serious dissent was expressed in
Charles A. Beard's great synthesis, published in 1927. Beard de-

plored the politics and other obsolete folkways surviving in the postwar era. But he found the center of current development, and the climax of his whole vast story, in the achievements of the machine age. Continuous invention was the hope of the future. Standardization had made possible not only better living for all but a more generous support for the life of the mind. Those who feared the machine were lumped together by Beard as "artists of a classical bent and . . . spectators of a soulful temper."[4] Lesser and more conventional historians usually struck the same note; and the textbooks of the period, if they ventured beyond Versailles, emerged into a few pages of peace and prosperity.[5]

Sociologists of the period, full of the élan of their new subject, exultant over the apparent defeat of religious obscurantism, were as optimistic as the businessmen and the historians, though for different reasons. Their New Era lay in the future rather than the present; its motivating force was not technology alone but the guiding social intelligence. This picture of the decade as a transitional age emerges most clearly from the sociological periodicals of the early twenties, where one finds at least four important assumptions. First, the scientific study of society is just coming into its own. Second, social scientists are now able to abandon sentiment, impressionism, and introspection and seek accurate information, especially quantitative information. Third, this new knowledge should be, and increasingly will be, the guide for practical statesmanship, replacing custom and tradition. Fourth, Utopia is consequently just around the corner. The present may look chaotic, but the new élite will be able to lead us fairly quickly out of the fog of dissolving tradition and toward the end of controversy and the reign of universal efficiency.[6]

To condense is always unfair, and it would be incorrect to assume that all social scientists in the twenties saw their role or their period this simply. Yet it is easy enough to find all these beliefs stated very positively in textbooks and even learned articles, with both the behaviorist dogmatism and the authoritarian implications full-blown. Part of the confidence of these prophets rested on real and important achievement by social scientists in the period, but those who had actually contributed the most new knowledge were sometimes less dogmatic than their colleagues. In *Middletown,* for instance, the social science interpretation of the twenties is buried in a mass of scrupulously collected facts, but it is there. At certain points in describing the decline of labor unionism or the standardization of leisure the authors seem to be deploring changes that have taken place since 1890. Yet in their conclusion they trace the tensions of Middletown to the lag of habits and institutions behind technological progress. Individual child-training, religion, and the use of patriotic symbols represent the past, while the future is represented by whatever is thoroughly secular and collective, particularly in the community's work life. The town has tended to meet its crises by invoking tradition in defense of established institutions. Their whole investigation, the Lynds conclude, suggests instead "the possible utility of a deeper-cutting procedure that would involve a re-examination of the institutions themselves."[7]

The typical economic thought of the twenties, while it avoided Utopian extremes, shared with the other social sciences an unlimited confidence in the present possibilities of fact-finding and saw in the collection and use of statistics much of the promise and meaning of the era. In his brilliant concluding summary of *Recent Economic Trends,* Professor Wesley C. Mitchell, for instance, found the main explanation for the progress of 1922-28 in the new application of intelligence to business, government, and trade-union administration.[8]

The third contemporary interpretation of the period, that offered by its literary intellectuals, differed sharply from the other two. Completely repudiating the optimism of the businessmen, it agreed with the social scientists only in its occasional praise of the liberated intelligence. For the most part, as we are all continually reminded, the writers and artists of the twenties saw their age as one of decline.

The most publicized group of pessimists was that typified by Harold Stearns and his colleagues of 1922, who, with their many successors, left an enduring picture of a barren, neurotic, Babbitt-ridden society. These critics have drawn a lot of patriotic fire, and indeed some of them are sitting ducks. They were often, though not always, facile, unoriginal, and ignorant. They seldom made clear the standards by which they found American society so lacking. Yet their lament is never altogether absurd or capricious. If one studies the civilization they saw around them through its press, one hardly finds it a model of ripeness or serenity. The fact remains, for historians to deal with, that American civilization in the twenties presented to many of its most sensitive and some of its most gifted members only an ugly and hostile face.

A more thoughtful and sadder group of writers than most of the young Babbitt-beaters traced their own real malaise not to the inadequacies of America but to the breakdown of the entire Western civilization. The New Humanists had long been deploring the decline of literary and moral discipline. At the opposite extreme in taste the up-to-date followers of Spengler agreed that decay impended. Joseph Wood Krutch in 1929 described the failure first of religion and then of the religion of science to give life meaning: "Both our practical morality and our emotional lives are adjusted to a world which no longer exists. . . . There impends for the human spirit either extinction or a readjustment more stupendous than any made before."[9]

Many accepted this statement of the alternatives, and chose according to their natures. Walter Lippmann, who had played some part in the confident prewar attack on tradition and custom, chose the duty of reconstruction and published, in 1929, his earnest attempt to find a naturalist basis for traditional moral standards.[10] On the other hand, T. S. Eliot painted a savage and devastating picture of present civilization and left it to live in the world which Krutch thought no longer existent. As Eliot assumed the stature of a contemporary classic, his description of the Waste Land, the world of Sweeney and Prufrock, and also his path away from it, seriously influenced later conceptions of the period.

With the Depression, the twenties shot into the past with extraordinary suddenness. The conflicting pictures of the decade, rosy and deep black, changed sharply, though none disappeared. Of them all, it was the New Era point of view, the interpretation of the decade as the birth of a new and humane capitalism, that understandably suffered most. Ironically, the most plausible and heavily documented version of this description, and one of the most influential later, appeared only in 1932 when Adolf A. Berle, Jr., and Gardiner C. Means described the separation of management from ownership.[11] At the time, however, the economic order of the twenties was collapsing, and its harassed defenders retreated temporarily into the Republican last ditch.

The other optimistic vision of the decade, that of the social scientists, depended less directly on prosperity and in the thirties survived somewhat better, though it became difficult to see the preceding period as the triumphant application of social intelligence. It is a startling example of the prestige of the social science point of view in 1929 that a President should commission a group of social scientists to make a complete and semi-official portrait of a whole civilization. The fact that *Recent Social Trends* was not completed and published until 1932 probably accounts in part for its excellence; it is the most informative document of the twenties which we have and also a monument of the chastened social science of the thirties. The committee that wrote this survey still believed, as its chairman, Wesley Mitchell, had earlier, that much of the meaning of the twenties lay in the harnessing of social intelligence to collective tasks. Consciously and subtly, the various authors documented the contradiction between the period's individualistic slogans and its actual movement toward social and even governmental control.[12] Yet they were conscious throughout that all this had ended in economic depression.

Like the authors of *Middletown,* the committee found its synthetic principle in the doctrine that change proceeds at different rates in different areas. Again like the Lynds, it assumed that society's principal objective should be "the attainment of a situation in which economic, governmental, moral and cultural arrangements should not lag too far behind the advance of basic changes," and basic here means primarily technological.[13] Occasionally *Recent Social Trends* displays, as for instance in its chapters on the child and on education, a surviving trace of the easy authoritarianism of the preceding decade's social theorists, and occasional chapters refer in the early optimistic manner to the hope of solving all social problems through the new psychological knowledge.[14] But in most of this great work, and particularly in its brilliant introduction, the authors left behind the social-science utopianism of the early twenties. It would take an increasingly powerful effort of social intelligence to bring us into equilibrium. Moreover, this effort must be a subtle one; the committee took pains to state that it was "not unmindful of the fact that there are important elements in human life not easily stated in terms of efficiency, mechanization, institutions, rates of change or adaptations to change."[15] Therefore, what was called for was not a ruthless rejection of tradition but a re-examination leading to a restatement in terms of modern life. *Recent Social Trends* is in places a work of art as well as of social science, and it is one of the few books about the twenties that point the way toward a comprehensive understanding of the period.

The view of the previous decade presented in the thirties by most historians was far less subtle and complete. Instead of either a New Era, a liberation, or a slow scientific adaptation, the twenties became a deplorable interlude of reaction. This view, stated sometimes with qualifications and sometimes very baldly, has continued to dominate academic historical writing from the thirties almost until the present.

Most of the historians who were publishing in the thirties had received their training in the Progressive Era. Many had been deeply influenced by Frederick Jackson Turner, and had tended to look for their synthesis not to the decline of Europe but to the expansion of America. Though the Turner doctrine can be turned to pessimistic uses, Turner himself in the twenties prophesied that social intelligence would find a substitute for the disappearing force of free land.[16] As this suggests, the outlook of John Dewey pervaded much of historical writing as it did the work of social scientists. Yet historians still tended to give most of their attention to politics. For these reasons, and because they shared the opinion of their readers, historians usually found the meaning of American history in the nineteenth-century growth of political and social democracy and the twentieth-century effort to adapt it to new conditions.

As we have seen, many of the historians actually writing in the twenties had not found their own period an interruption of this beneficent adaptation. The interruption had come in 1929 and then, after an interval of confusion and paralysis, Franklin D. Roosevelt had appealed for support partly in terms of the progressive view of history. Roosevelt himself justified his program by pointing to the end of free land[17] and claimed the progressive succession from Theodore Roosevelt and Woodrow Wilson, his cousin and his former chief. Few historians were disposed to deny his claim, and accepting it made the twenties an unfortunate interregnum, sometimes covered by a chapter called "The Age of the Golden Calf," or "Political Decadence," or even "A Mad Decade."[18]

This does not mean that an emphasis on the political conservatism of the decade, or a hostile criticism of the Harding-Coolidge policies, is in itself a distortion. Yet stubborn standpattism was only one ingredient in a varied picture. It is not history to make the twenties, as some of the briefer historical treatments do, merely a contrasting background for the New Deal. Sometimes even prosperity—an important fact despite the exceptions—is belittled almost out of existence, the prophets of abundance are denied credit for good intentions, the approach of the Depression becomes something that nearly anybody could have foreseen, and the decade's many advances in science, social science, medicine, and even government are left out.[19]

While they deplored the businessmen and politicians of the twenties, the progressive historians of the thirties and later tended also to belittle the period's literary achievement. This negative judgment was sustained by a powerful writer, Vernon L. Parrington, himself a thorough and fervent exponent of the progressive interpretation of American history. In Parrington's last, fragmentary volume, published in 1930, he read the younger authors of the twenties out of the American tradition as "a group of youthful poseurs at the mercy of undigested reactions to Nietzsche, Butler, Dadaism, Vorticism, Socialism; overbalanced by changes in American critical and creative standards, and in love with copious vocabularies and callow emotions." "With the cynicism that came with postwar days," said Professor Parrington, "the democratic liberalism of 1917 was thrown away like an empty whiskey-flask."[20]

Though Parrington did not live to explain this rejection or treat it at length, he obviously believed that the liberal whisky was still there and still potent, and so, in the thirties and often since, have many of his readers. Some historians, understandably impressed by Parrington's great architectural achievement, willingly and specifically took over his literary judgments; others doubtless arrived at similar opinions independently.[21] For whatever reason, by the thirties the most widespread historical picture of the twenties was that of a sudden and temporary repudiation of the progressive tradition by reactionary politicians and also by frivolous or decadent littérateurs.

Some of the historians writing in the thirties, and far more of the literary critics, found their historical principle not in American progressivism but in Marxism. John Chamberlain demonstrated to his own temporary satisfaction the futility of the preceding Progressive Era, and Lewis Corey and others depicted the resultant triumph of monopoly capitalism, characterized by a false prosperity and leading inevitably to the Depression and (before 1935) the disguised fascism of the New Deal.[22] At their worst, and in most of their specifically historical writing, the Marxist writers seem now unbelievably crude and schematic. But the Marxist version of the twenties came not only from the pamphleteers but also from gifted literary artists. For many of the generation that grew up in the thirties the concept of the previous decade was strongly influenced by the work of John Dos Passos. His brilliant sketches of Woodrow Wilson, Henry Ford, Thorstein Veblen, and other giants, the post-war violence, the defeat of hopes, and the gradual inevitable corruption of the "big money" form a picture that is hard to forget—that Dos Passos himself in sackcloth and ashes is entirely unable to wipe out. Among the many critics and literary historians who were then Marxists, most of them dull and fashion-ridden, were a few writers of insight. It is still suggestive to see the literary rebellion of the twenties, through the 1935 eyes of Granville Hicks, as a reflection of the insecurity of the middle class.[23] Most of the rebellious writers *had* come from this class, and even from a particular segment of it that had lost prestige, and many of them had been self-conscious and worried about this origin.

Sometimes, despite their basic differences, the Marxist writers agreed in part in the thirties with the progressive historians. Often, however, the literary Marxists made a different combination. Starting in the twenties as rebels in the name of art, they had found their esthetic distaste for capitalism confirmed by prophecies of its inevitable doom. The resultant mixture of individualist rebellion and socialist revolution was unstable and short-lived, but in the thirties powerful. Edmund Wilson describes the representative mood, and the resultant attitude toward the twenties: "To the writers and artists of my generation who had grown up in the shadow of the Big Business era and had always resented its barbarism, its crowding-out of everything they cared about, these Depression years were not depressing but stimulating. One couldn't help being exhilarated at the sudden unexpected collapse of that stupid gigantic fraud."[24]

One other and opposite group of writers in the thirties contributed to the previous decade's bad press. This was the varied group stemming from T. S. Eliot's neo-classical essays and I. A. Richards's effort at a scientific criticism that came to be known as "the New Critics." This school of writers could almost be defined as a counter revolution against the individualist rebellion of the twenties, in which some of them, not surprisingly, had themselves played a part. Some of the New Critics called for a revival of the Catholic, or Anglo-Catholic, or humanist, or southern tradition; others hoped to find a new credo in literature itself. They agreed only in valuing such qualities as complexity, tension, and intellectual strictness. In the thirties, despite the noise made by opposite groups, it was the New Critics who were moving quietly toward a position of dominance in criticism and in the college teaching of literature; a position they clearly hold today.

Like their enemies, the Marxists and progressives, the New Critics found little to praise in the twenties. To begin with, they stoutly rejected any tendency to measure the progress of civilization in terms of technology or standard of living. Thus they saw both the business civilization of the New Era and the opposing humanitarian progressivism as two variants of the same shallow materialism.[25] To them the social science Utopias forecast in the twenties were merely a repulsive climax to current tendencies. Allen Tate, for instance, associated social science not only with innocent barbarism but with the current triumph of the total state: "What we thought was to be a conditioning process in favor of a state planned by Teachers College of Columbia University will be a conditioning equally useful for Plato's tyrant state. . . . The point of view that I am sketching here looks upon the rise of the social sciences and their influence in education, from Comtism to Deweyism, as a powerful aid to the coming of the slave society." Looking back at the previous period, Tate remembered sadly "How many young innocent men—myself among them—thought, in 1924, that laboratory jargon meant laboratory demonstration."[26]

Most of the New Critics rejected the rebellious literature of the twenties as completely as they did the business civilizations of the era. Exceptions had to be made, of course, for the more careful and rigorous poets—Marianne Moore, Eliot, sometimes Ezra Pound. The abler of the New Critics realized, as some moralists did not, that the writers of the twenties expressed, rather than caused, the disintegration of tradition which they deplored. Some of them were able to admire men like Ernest Hemingway and Hart Crane who bravely tried to give literary form to moral and intellectual disorder. But the general direction of the literature of the decade was, they agreed, disintegration.[27]

Progressives, Marxists, and neo-classicists all found the twenties deplorable, yet in writers from all these camps, and in others who wrote in the thirties, a note of nostalgia often broke through the sermon. Frivolous, antisocial, and decadent as the literature of the twenties seemed, it had to be conceded the somewhat contradictory qualities of freshness and excitement. And nostalgia, in the thirties, extended beyond the previous decade's literature to its manners and customs. In 1931 Frederick L. Allen performed a remarkable feat of impressionist recall of the period just over, and in 1935 Mark Sullivan brought back vividly its clothes and songs and sensations.[28] Already in the work of these two excellent reporters, and later in the versions of a number of minor and more sentimental merchants of nostalgia, the twenties appeared strange, fantastic, and appealing. They appealed with particular strength to those who did not remember them; it was the peculiar feat of these reporters to fill the new generation with nostalgia for scenes they had not seen. For the college student of the next decade, if the twenties was one half the betrayal of progress, the other half was the jazz age. Irresponsibility, to the solemn and uneasy thirties, was both deplorable and attractive.

This paradoxical attitude toward the twenties continued and the paradox sharpened in the next period. In the dramatic and tragic days of World War II, few found much to admire in the age of Ford and Coolidge. James Burnham, combining Berle and Means's data on the separation of ownership and control with an apocalyptic vision of the rise of the total state, made the New Era into the beginning of the "Managerial Revolution."[29] To the F.D.R. liberals, who already blamed the twenties for abandoning progressivism, the period's major crime was now its rejection of the Wilsonian international program. Teachers worried whether the earlier postwar disillusion, which they had helped to propagate, would make it impossible to revive a fighting spirit—a worry which proved unnecessary and perhaps a little conceited. Editorial writers wondered whether the country would again fail in its responsibilities after the war. Above all, those who responded most generously to the call for the defense of Western culture feared that the literary rebels of the twenties had done great, even disastrous, damage to the nation's morale.

Even before the war broke out, Walter Lippmann was concerned about the lack of fighting convictions among civilized men and blamed, in part, the rejection of tradition in which he had long ago taken part. Archibald MacLeish blamed both the artists and the scholars of the previous period for their different kinds of detachment. Van Wyck Brooks, looking back at the writers who had answered his own summons for a new literature, found that they differed from all previous writers in one striking way: they had ceased to be "voices of the people."[30] "How could a world," he wondered, "that was sapped by these negative feelings resist the triumphant advance of evil."[31]

This high estimate of the power and responsibility of literature seemed to be shared by Bernard DeVoto, though he took writers to task for making literature the measure of life. Writers of the "Age of Ignominy" had condemned their period partly out of sheer ignorance. In his eagerness to demonstrate this DeVoto revived, earlier than many, some of the New Era interpretation of the twenties. "What truly was bankrupt was not American civilization but the literary way of thinking about it." Actually, "The nation that came out of the war into the 1920's was . . . the most cheerful and energetic society in the world."[32] A true picture of it would have emphasized its achievements in education, medicine, humanitarian improvement, and the writing of local history.

MacLeish, Brooks, DeVoto, and others condemned the writers of the twenties for damaging the nation's fighting morale, and strangely enough, Charles and Mary Beard, writing in 1942 of the American Spirit, made the same charges from an isolationist point of view. For the Beards, American cynicism had come from Europe: "In the tempers and moods fostered by foreign criticisms and by American weakness displayed in reactions to the impacts, multitudes of young men and women were brought to such a plight that they derided the whole American scene."[33]

All these works, including in part that of the Beards (which was not one of the major productions of these great historians), were wartime pamphlets rather than history. None of them offered a halfway satisfactory explanation of the alienation they discussed, which was certainly a more important phenomenon than the inadequacy of a few individuals. Yet one thing the wartime writers said was true and worth saying, that in the twenties a deep chasm had opened between the views of life of most writers and their fellow citizens. Perhaps the importance of this fact could not be emotionally grasped until the years when DeVoto heard Ezra Pound on the Italian radio.

Yet, even in wartime, and for some perhaps especially in wartime, the freedom and creativity and even the irresponsibility of the previous generation of writers had a paradoxical attraction. Alfred Kazin's admirable and by no means uncritical chapters on the period, which appeared in 1942, were called "The Great Liberation (1918-1929)."[34] And the paradox seemed to reach its most acute form in DeVoto himself. In the same short volume the literature of the twenties was "debilitated, capricious, querulous, and irrelevant" and yet the decade was "one of the great periods of American literature, and probably the most colorful, vigorous, and exciting period." It was a literature that was "not . . . functional in American life," but "idle, dilettante, flippant, and intellectually sterile,"and yet one which had "achieved something like a charter of liberties for American writers."[35]

In the 1950s, as in other periods, it is dangerous to equate the latest insights with truth. Yet it is hard not to conclude that now, in the second postwar period, some writers are converging from various directions toward a better understanding of the twenties. For one thing, the decade is longer past and it is no longer acutely necessary to break with its viewpoint. Fairly recently the twenties have come to be a fair field for the dissertation and the monograph, which bring at least a different kind of knowledge. One survivor of the period says that instead of being revived, it is being excavated like a ruin, and another complains that he and his friends are already being preserved in complete bibliographies while yet, as far as they can tell, alive.[36]

Disapproval and nostalgia, of course, remain. Editorials worry about the effect on Europe of the vogue there of the literature of the twenties. Professor Howard Mumford Jones has continued something like DeVoto's charges in more analytic tones, accusing the postwar writers both of brilliance and of detachment amounting to solipsism.[37] The choice of Scott Fitzgerald for revival and in some quarters canonization indicates the perverse attraction which self-destruction seems to hold for our period. Budd Schulberg's novel specifically contrasts a romantic and defeated alcoholic writer of the twenties with a crass, earnest young radical of the thirties to the latter's obvious disadvantage.[38]

In general, however, literary opinion seems to have gone beyond both nostalgia and reproof into a more mature and solidly based appreciation of the achievements of this era now so safely in the past. To many, the apparent sterility of the present literary scene furnishes a depressing contrast. Whatever else they rejected, writers of the twenties took their writing seriously, and, as Cowley has pointed out, publishers made it possible for them to do so.[39] Professor Frederick J. Hoffman in the most thorough of many recent accounts finds the period's literature full of daring, variety, and technical brilliance. This estimate by now represents more than a cult; it is an accepted consensus.[40]

One achievement of the twenties which has received only a little specific comment is nevertheless widely recognized today. The period of alienation and exile gave rise, curiously enough, to a thorough, rich, and continuing inquiry into the whole American past. The sources of this inward turn are as complicated as the decade itself. Many of the major historians who wrote then, including Parrington, Beard, Carl Becker, and Arthur M. Schlesinger, Sr., belong to the group that always found its major synthesis in the course of democratic progress. But others turned to the past with Van Wyck Brooks, partly in a spirit of cultural nationalism, to destroy the English and Anglophile genteel tradition and replace it with something native. Still others went first through a phase of violent rejection of American culture and then, finding Europe essentially unavailable as a substitute, returned to look desperately for roots at home. By the forties and fifties it was possible to see the lines converging in a cultural history which, at its best, could be critical, conscious of irony and failure, and yet, in a meaningful and necessary way, patriotic.[41]

With the literature and historical research of the twenties, its economic achievement, once overvalued and then rated too low, has again turned the corner into a rising market. In the years of the Marshall Plan, when American capitalism was called on to shoulder an immense burden, it was hard to think of it as a failure and a mistake. And in the still rising prosperity of the Eisenhower period, far more widespread and soundly based than that under Coolidge but inevitably reminiscent, a reassessment of the earlier period was natural enough.

Part of the reassessment arose from the increasing complexity of economics and the development of a new economic history.

Beginning about 1940, a number of economists and historians had demanded that American economic history separate itself from the political framework and give more attention to such matters as real wages and volume of production, and somewhat less to labor organization and the political struggles between farmers and merchants.[42] Even earlier, the business historians had been asking for a more analytic and less emotional approach to the history of management.[43] By the forties, it was impossible for an informed historian to duplicate the sweeping judgments about the boom and crash that had been easy ten years earlier. In 1947 George Soule, in his detailed economic history of the twenties, concluded perhaps rather to his own surprise that the rich grew richer without the poor growing poorer, that new amenities became available on a scale impossible to ignore, and that no measures then available would certainly have prevented the crash.[44] Most of the more recent economic history textbooks seem either to suggest a similar assessment or to avoid passing judgment altogether. Even the economic foreign policy of the twenties, long a favorite target of liberal historians, has been presented by Herbert Feis as a well-intentioned though ineffective forerunner of Point Four.[45] In 1955 John K. Galbraith, even in a book on the "Great Crash," took historians mildly to task for underrating what was good in the Coolidge era, and unfairly blaming Coolidge himself for a failure of prophecy.[46]

Such opposite kinds of writers as Peter Drucker, Frederick L. Allen, and the editors of *Fortune* have argued, without special reference to the twenties, that American capitalism since about the turn of the century has been evolving into a new kind of democratic and humane economic order.[47] Most recently David M. Potter concludes that we have always been the "People of Plenty" and that this fact, more than the frontier or political freedom, has shaped our mores.[48] Professor Potter, more sophisticated than earlier prophets of abundance, has learned from the social scientists that a country has to pay for production in competitive strain, and perhaps later for security in loss of mobility. Yet his perspective, like that deriving from our whole political and economic climate, shifts the meaning of the earlier prosperity era. If productivity holds much of the meaning of American history, it is the Depression and not the twenties that marks the interruption in a steady development. The New Era represents at worst a promising try at a new economy, a chapter in a book with a happy ending.

There is much in this reassessment that is invigorating, especially in a period when the leftist clichés are the tiredest of all. Yet several cautions are in order. Historians must remember, first, that the early 1880s and the 1920s and the 1950s are different and separated periods of prosperity, no matter how similar; second, that the depressions, even if in the long run temporary interruptions, did not look that way to their victims; and third, that even complete economic success does not, either now or for the twenties, refute all criticisms of American culture.

There is little danger that we will altogether forget this last caution. While some contemporary writers present a view of our recent history that emphasizes economic success, to another group such success is not so much false as irrelevant. The anti-optimists today are not rebels but traditionalists, a group that can be lumped together as anti-materialist conservatives. Some of these derive from and continue the new criticism, others reflect the revival of theology, and still others rely partly on new scientific theory. All have been led or forced, during the recent era of world catastrophe, to place their trust not in secular progress but primarily in moral and religious tradition, and from this standpoint the twenties are difficult to rehabilitate.

Joseph Wood Krutch has devoted a volume to repudiating the mechanistic determinism he voiced so powerfully in 1930, and Walter Lippmann has even more specifically repudiated his early relativism. In 1955 Lippmann concluded that the whole debacle in international politics, starting in 1917 and continuing through and after Versailles, resulted primarily from "the growing incapacity of the large majority of the democratic peoples to believe in intangible realities," specifically in a transcendent, universally valid, natural law.[49]

Many powerful contemporary writers agree with Lippmann not only in his diagnosis of the trouble but in his fixing the responsibility for breakdown in the 1920s. Some of these, however, find in the decade enough just men to save it from complete condemnation. Russell Kirk, for instance, resurrects the New Humanists and marvels that "these years of vulgarity and presumption" produced the coming of an age of American conservatism in a group of thinkers who struggled against "the vertiginous social current of the Harding and Coolidge and Hoover years."[50] (It marks perhaps the high point in this reassessment to make Coolidge, rather than Freud or Einstein, a symbol of vertigo.) A more subtle conservative and antimaterialist finds in the literary rebels the saving remnant. In his curious, dogmatic, but occasionally suggestive *Yankees and God*, Chard Powers Smith suggests that the young iconoclasts of the twenties were really the last, or next-to-the-last, wave of Puritanism, despite their use of the term Puritan as the ultimate of abuse.[51] This apparently bizarre thesis is really neither absurd nor entirely original. Perry Miller in 1950 gave the rebels of the twenties a similarly respectable pedigree when he compared them to the transcendentalists. Both of these movements spoke for the spirit against the rule of things, and both, said Professor Miller, belonged in a series of "revolts by the youth of America against American philistinism."[52] One can go a very little further and agree with Mr. Smith that both are basically Protestant; it is not hard to recognize in the young intellectuals of the twenties together with their iconoclasm a tortured uneasiness, a conscious responsibility for the faults of the era that are suggestive of a long heritage.

In the 1950s, then, the familiar division continued. Spokesmen of the New Era rehabilitated the twenties by using one set of standards while anti-materialists blamed or praised them according to another. At the same time, however, a number of scholars of varying views were reaching toward an understanding of such paradoxes by treating the twenties as a period of profound social change. Most of these students derived their insights to some extent from the sociologists, and it is interesting that some of the gloomiest insights stem today from this once exuberant science. David Riesman's strikingly influential vision of the shift from inner-direction to other-direction is not strictly dated by its creator, but it often seems to be a description of the end of the genteel tradition and the birth of the New Era, the defeat of Wilsonian moralism and the victory of the Babbitts.[53] In different terms and with a more clearly stated value judgment, C. Wright Mills has documented the rise of a regimented, rootless, and docile new middle class to the arbitral position in American society.[54] The increase of the white-collar salariat and its implications extended before and after the twenties but went especially fast in that period, as the authors of *Recent Social Trends,* among others, pointed out. Samuel Lubell and others have seen another social change in the twenties, the beginning of the coming-of-age of the new immigration.[55] Drawing together Lubell's interpretation and Mills's, Richard Hofstadter emphasizes the "Status Revolution" as a main event of the period about the turn of the century.[56] The Protestant upper middle class, long a semi-aristocracy with a monopoly on advanced education, had declined, and so had the independent farmers. In their places other groups had grown and gained some power—the new middle class, the ethnic minorities, and labor. All these processes of change had, by the twenties, proceeded a long way, and all were continuing and accelerating, with the partial exception of the rise of labor. Surely this social upheaval, impossible to see clearly until our own time, has considerable meaning for the intellectual history of the twenties as for its politics, for the collapse, that is, of a long-frayed moral and literary tradition.

The nearest we can come to summarizing or explaining the shifting opinions of the twenties may well be to see the period in some such terms as these, and to see it as a disintegration. There is certainly nothing original about such a conclusion, but perhaps we are now in a position to give disintegration a fuller and more various meaning. The twenties were a period in which common values and common beliefs were replaced by separate and conflicting loyalties. One or another of the standards arising from the age itself has been used by each of its historians ever since. This is what has made their judgments so conflicting, so emotional, so severally valid and collectively confusing. It is equally true and equally partial to talk about the rising standard of living and the falling standard of political morality, the freshness and individuality of literature and the menace of conformity, the exuberance of manufacturers or social scientists and the despair of traditional philosophers. Somehow, we must learn to write history that includes all these, and the first step is to understand the decade when the fragmentation first became deep and obvious.

At least two recent writers are useful to those who want to look at the twenties from this point of view. One is Lionel Trilling, who deplores and analyzes the split between liberalism and the imagination, between the values we take for granted as socially desirable and those that have now the power to move us in art, between collective welfare and individual dignity.[57] What is lacking, says Trilling, and what has been lacking specifically since the twenties, is a view of the world, in his word a faith though not necessarily a religion, that will give meaning both to society and to art, to progress and to tragedy. Professor Henry Nash Smith in a recent address has sketched, somewhat similarly, two diametrically opposite points of view which, he says, have divided our culture since 1910.[58] One he calls the realistic-progressive view and the other the counter-enlightenment; one takes for its standards measurable welfare and humanitarian progress and equality; the other values only the individual imagination, nourished on tradition, holding out desperately against a mechanized culture, and accepting if necessary alienation and despair as the price of its survival.

The conflict of values that culminated, for it certainly did not begin, in the twenties was more than two-sided, and neither of these two critics has completely explored it. But they have indicated the right starting point. The way to understand our recent cultural history is to understand why and how its exponents fail to agree.

How can historians proceed further along this path? First, it hardly needs saying that to understand the twenties better we must make use of techniques drawn from various fields. The most important developments in the decade did not take place in the realms of politics, or economics, or literature, or science alone, but in all these areas and the relation, or lack of relation, among them. If one uses one kind of sources one will inevitably emerge with one point of view, which will be inadequate to understand the others.

Second, it seems clear that one cannot say much about the twenties as a disintegration or revolution without giving more attention to the old regime, the presumed prewar agreement. There seems to have been a greater degree of unity in American culture before 1917 or perhaps 1910, but a description of it is not easy and a casual reference to the genteel tradition or the cultural inheritance will not suffice. Immediately prewar America must be newly explored. We must look not so much at its articulate political or philosophical beliefs and more at its inarticulate assumptions—assumptions in such areas as morality, politics, class and race relations, popular art and literature, and family life. In short, we must concentrate on what Tocqueville would have called its manners. We are now, perhaps, in a position at least to undertake this recapture in an impartial mood. In 1956 we do not need to lament or rejoice at the destruction of the America of 1914; it is nearly as far off as Greece or Rome, and as inevitably a part of us.

Third, we must try to look at the succeeding disintegration, the revolution of the twenties, with a similar absence of passion. The literary scoffers who have been so thoroughly scolded were not, after all, the only rebels. The prophets of mechanization and welfare, the Fords and Edisons who scorned history and tradition, were equally revolutionary. Most revolutionary of all, perhaps, were the prophets of psychology and social science, with their brand new societies full of brand new human beings.

Finally, if we can really look back on this revolutionary decade from a perspective which has the advantage of thirty years of continuing revolution, we may be able to see which of the separate movements of the twenties has lasted best, and whether any of them are beginning to come together. Are there really in this decade of novelty beginnings as well as ends? Is it possible by now really to glimpse what so many have announced: the beginnings of a new period of American history and even of a new civilization?

Notes

1. Thomas N. Carver, *The Present Economic Revolution in the United States* (Boston, 1926), is perhaps the most effective single presentation of this common version of the period.
2. Preston W. Slosson, *The Great Crusade and After* (New York, 1930), 729.
3. James C. Malin, *The United States after the World War* (Boston, 1930), 530-43. Like some of the social scientists discussed below, Malin thought that "It is possible that in the long run the changes even extended effective governmental regulative powers, although critics of the new policies held the opposite view" (p. 540).
4. Charles A. and Mary R. Beard, *The Rise of American Civilization* (2 vols., New York, 1927), II, 729.
5. Paul L. Haworth, *The United States in Our Own Times, 1865-1920* (New York, 1920), called his last chapter "A Golden Age in History," and left both title and contents nearly unchanged in his editions of 1924, 1925, and 1931. More temperately, Samuel E. Forman, *Our Republic* (Rev. ed., New York, 1929), 881, balanced "stupendous productivity" against such blemishes as technological unemployment and concluded that the country was "sound at the core."

6. For optimism about the prospects of social science, see for instance Emory S. Bogardus's preface to Elmer S. Nelson, Charles E. Martin, and William H. George, *Outlines of the Social Sciences* (Los Angeles, 1923), xvii-xx. For a strong statement about the role of social scientists in correcting all existing abuses, see John Candler Cobb, "The Social Sciences," *American Journal of Sociology* (Chicago), XXXI (May, 1926), 721. An unusually strong statement of the necessity for the well-informed to control society is that by the historian of the social sciences, Harry Elmer Barnes, "History and Social Intelligence," *Journal of Social Forces* (Chapel Hill), II (November, 1923), 151-64.

7. Robert S. and Helen M. Lynd, *Middletown* (New York, 1929), 502.

8. President's Conference on Unemployment, *Recent Economic Changes* (New York, 1929), 862.

9. Joseph W. Krutch, *The Modern Temper* (New York, 1929), 26.

10. Walter Lippmann, *A Preface to Morals* (New York, 1929).

11. Adolf A. Berle, Jr., and Gardiner C. Means, *The Modern Corporation and Private Property* (New York, 1932).

12. President's Research Committee on Social Trends, *Recent Social Trends* (2 vols., New York, 1933). This is a main theme of Chapters 23 to 29, II, 1168-1541.

13. *Ibid.,* I, lxxiv.

14. *Ibid.,* II, 1185.

15. *Ibid.,* I, lxxv.

16. See his statement of 1924, quoted in Henry Nash Smith, *Virgin Land* (Cambridge, 1950), 258-59.

17. See his famous Commonwealth Club Address, Samuel I. Rosenman (ed.), *The Public Papers and Addresses of Franklin D. Roosevelt* (13 vols., New York, 1938-1950), I, 742-56.

18. The first two of these occur in Dwight L. Dumond, *Roosevelt to Roosevelt* (New York, 1937), the general title of which indicates its outspoken loyalties; the last in James Truslow Adams, *The March of Democracy* (2 vols., New York, 1933). Adams's best-selling *Epic of America* (Boston, 1931) contains one of the most complete indictments of all aspects of the culture of the twenties. The above generalizations about American historians do not, however, apply fully to Adams, whose ideas are somewhat atypical. His dislike of the decade's culture was expressed early in his *Our Business Civilization* (New York, 1929), which repeats many of the criticisms made by the literary anti-conformists.

19. Fred A. Shannon, *America's Economic Growth* (Rev. ed., New York,

1940), describes the economic policies of the period thus: "It was in this atmosphere of rapacity and high-pressured seduction that governments reverted to *laissez faire* policy, contorted to mean government assistance to business" (p. 585), and refers to the "fools' paradise" and the "years of paper prosperity" of the period (pp. 701, 727). A later judgment is that of Henry B. Parkes, *Recent America* (New York, 1946), that "There was probably more materialism, more illiberality, and more cynicism than ever before in American history" (p. 464). One can think at least of close contenders to some of these titles.

20. Vernon L. Parrington, *Main Currents in American Thought* (3 vols., New York, 1927-1930), Vol. III, *The Beginnings of Critical Realism in America*, 385-86, 412.

21. An example is Louis M. Hacker, *American Problems of Today* (New York, 1938), which quotes and cites Parrington's judgments liberally (e.g., p. 165). A historian who states his admiration of Parrington very strongly in our own time is Henry Steele Commager, *The American Mind* (New Haven, 1950), 445.

22. Lewis Corey, *The Decline of American Capitalism* (New York, 1934), and *The Crisis of the Middle Class* (New York, 1935). An example of Marxist interpretation at its simplest is Bruce Minton and John Stewart, *The Fat Years and the Lean* (New York, 1940).

23. Granville Hicks, *The Great Tradition* (New York, 1935), 215.

24. Edmund Wilson, "The Literary Consequences of the Crash" (first published in 1932), in Wilson, *The Shores of Light* (New York, 1952), 409.

25. This radical separation of material and spiritual values may be found in Eliot's essays in the early twenties and is strongly stated in John Crowe Ransom, "Flux and Blur in Contemporary Art," *Sewanee Review* (Sewanee, Tenn.), XXXVII (July, 1929), 353-66. It was in the thirties, however, that the New Critic movement drew together as a school. As early as 1931, Max Eastman acutely pointed out that this radical dualism was a curious attitude in those who wanted to restore the unity of Western cultural tradition. Eastman, *The Literary Mind* (New York, 1931).

26. Allen Tate, *Reason in Madness* (New York, 1935), 7, 11.

27. A good sample of the attitude of the New Critics toward the twenties, conveying both the acuteness and the dogmatism of the movement, is Richard P. Blackmur, "Notes on E. E. Cummings's Language," published in *Hound and Horn* (Portland, Me.), in 1931, and reprinted in Morton D. Zabel's very helpful anthology, *Literary*

Opinion in America (New York, 1937; rev. ed., 1951), 296-314. A typical verdict from an atypical critic is that of Yvor Winters: "During the second and third decades of the twentieth century, the chief poetic talent of the United States took certain new directions, directions that appear to me in the main regrettable. The writers between Robinson and Frost, on the one hand, and Allen Tate and Howard Baker on the other, who remained relatively traditional in manner were with few exceptions minor or negligible; the more interesting writers . . . were misguided." Winters, *Primitivism and Decadence* (New York, 1937), 15. A little later Randall Jarrell acutely suggested, from a New Critic point of view, the similarity between the period's rebels and its dominant tendencies: "How much the modernist poets disliked their society, and how much they resembled it! How often they contradicted its letter and duplicated its spirit! They rushed, side by side with their society, to the limits of all tendencies." Jarrell, "The End of the Line" (first published in 1942), in Zabel, *Literary Opinion in America* (rev. ed.), 742-48.

28. Frederick L. Allen, *Only Yesterday* (New York, 1931); Mark Sullivan, *Our Times: The United States, 1900-1925* (6 vols., New York, 1926-35), Vol. VI, *The Twenties*. A later sensational and amusing treatment of some aspects of the decade is Laurence Greene, *The Era of Wonderful Nonsense* (Indianapolis, 1939).

29. James Burnham, *The Managerial Revolution* (New York, 1941).

30. Walter Lippmann, *The Good Society* (New York, 1937); Archibald MacLeish, *The Irresponsibles* (New York, 1940); Van Wyck Brooks, *The Opinions of Oliver Allston* (New York, 1941). "Allston" condemns the rebellious poets and novelists of the twenties and, even more vigorously, their opponents the New Critics (as "coterie writers," pp. 241 ff.). He rejects the "excuses" characteristic of the postwar authors and insists that the trouble is not relativity, mechanization, etc., but the emotional inadequacy of the writers themselves (pp. 249-50).

31. Brooks, *Opinions of Allston,* 205. The opinions quoted are those of "Allston," Brooks's thinly disguised fictional counterpart.

32. Bernard DeVoto, *The Literary Fallacy* (Boston, 1944), 123, 162.

33. Charles A. and Mary R. Beard, *The American Spirit* (New York, 1942), 484. For the Beards, as for many other cultural historians, "The American Philosophy" is that of John Dewey (p. 665). This version of American intellectual history seems to need considerable qualification.

34. Alfred Kazin, *On Native Grounds* (New York, 1942), 187.

35. DeVoto, *Literary Fallacy,* 13, 15, 165-66, 169.

36. Malcolm Cowley, *The Literary Situation* (New York, 1954), 3; Edmund Wilson, "Thoughts on Being Bibliographed," *Princeton University Library Chronicle* (Princeton), V (February, 1944), 51-61.

37. Howard M. Jones, *The Bright Medusa* (Urbana, Ill., 1952). Jones analyzes with considerable success both the attraction of the twenties and what he sees as their characteristic faults.

38. Budd Schulberg, *The Disenchanted* (New York, 1950). As some reviewers pointed out, Schulberg is not sure whether he more admires or pities his major character, clearly modeled on Fitzgerald. The Fitzgerald revival reached its greatest extent with the discussions arising out of Arthur Mizener's biography, *The Far Side of Paradise* (New York, 1950).

39. Malcolm Cowley, "How Writers Lived," Robert E. Spiller *et al., Literary History of the United States* (Rev. ed., New York, 1953), 1263-72.

40. Frederick J. Hoffman, *The Twenties: American Writing in the Postwar Decade* (New York, 1955). Another estimate that emphasizes the same qualities is John K. Hutchens in his preface to his anthology, *The Twenties* (Philadelphia, 1952), 11-34.

    A critical but high estimate from the point of view of a present-day novelist is that of James A. Michener, "The Conscience of the Contemporary Novel," Lewis Mumford *et al., The Arts in Renewal* (Philadelphia, 1951), 107-40.

41. For a helpful analysis of the development of American literary studies, see Howard M. Jones, *The Theory of American Literature* (Ithaca, 1948). A contemporary document which brings out the various approaches of the twenties to the American past is Norman Foerster (ed.), *The Reinterpretation of American Literature* (New York, 1928). An extreme example of the tendency today to credit the twenties with a major accomplishment in this respect is Malcolm Cowley's dictum that "Perhaps the greatest creative work of the last three decades in this country has not been any novel or poem or drama of our time . . . perhaps it has been the critical rediscovery and reinterpretation of Melville's *Moby Dick* and its promotion step by step to the position of national epic." Cowley, "The Literary Situation: 1953," *Perspectives USA* (New York), No. 5 (Fall, 1953), 5-13. This promotion began in the twenties and owes much to the outlook of that decade.

42. A most valuable account of the beginnings of this movement is Herbert Heaton, "Recent Developments in Economic History," *American Historical Review* (New York), XLVII (July, 1942), 727-46. But note that the results seem barely yet apparent to Mr. Heaton in a review of four economic histories in *Mississippi Valley Historical Review* XXXVIII (December, 1951), 556-61.

43. Norman S. B. Gras, *Business and Capitalism* (New York, 1939), states the point of view of the business historians. The genesis and progress of the movement are excellently described in Henrietta M. Larson's introduction to her *Guide to Business History* (Cambridge, 1948), 3-37.

44. George Soule, *Prosperity Decade* (New York, 1947), especially p. 335.

45. Herbert Feis, *The Diplomacy of the Dollar: First Era, 1919-1932* (Baltimore, 1950). I have not mentioned among the recent optimistic historians of the twenties Professor Frederic L. Paxson, whose detailed volume on the period is, by the author's design, as lacking in interpretative comment as it is possible for a book to be. Paxson's occasional generalizations, however, indicate that he did not regard the twenties as an interruption in the readjustment of the federal government to the facts of a changing life, "and even that a new pattern was developing in American society, a pattern which meant for many Americans a more open future." Frederic L. Paxson, *American Democracy and the World War* (3 vols., Boston and Berkeley, 1936-48), Vol. III, *Postwar Years: Normalcy, 1918-1923*, Introduction, 2.

46. John K. Galbraith, *The Great Crash, 1929* (Boston, 1955), 608.

47. Peter F. Drucker, *The New Society* (New York, 1949); The Editors of *Fortune, U. S. A.: The Permanent Revolution* (New York, 1951); Frederick L. Allen, *The Big Change* (New York, 1952). In the last two of these it is not altogether clear whether the twenties are a part of the fortunate development or a break in it.

48. David M. Potter, *People of Plenty* (Chicago, 1954).

49. Joseph W. Krutch, *The Measure of Man* (New York, 1953); Walter Lippmann, *The Public Philosophy* (Boston, 1955), 55.

50. Russell Kirk, *The Conservative Mind* (Chicago, 1953), 362-63.

51. Chard Powers Smith, *Yankees and God* (New York, 1954), 451-59.

52. Perry Miller, *The Transcendentalists* (Cambridge, 1950), 8, 14-15.

53. David Riesman, *The Lonely Crowd* (New Haven, 1950).

54. C. Wright Mills, *White Collar* (New York, 1951).

55. Samuel Lubell, *The Future of American Politics* (New York, 1953), 34-41.
56. Richard Hofstadter, *The Age of Reform* (New York, 1955), 131-72.
57. Lionel Trilling, *The Liberal Imagination* (Pocket ed., New York, 1950 [first published, 1948]), especially pp. 97-106, 245-87.
58. Henry Nash Smith, "The Reconstruction of Literary Values in the United States, 1900-1950" (unpublished manuscript, 1952).

· 3 ·

The Other
American Tradition

This was an inaugural lecture, delivered at the Universities of Liège, Brussels, and Ghent, at which I was a Fulbright lecturer in 1959-60. It was the custom for the University dignitaries, a large number of faculty members, and sometimes representatives of the American Embassy to attend Fulbright inaugural lectures. Thus the lecture was prepared for an audience of highly intelligent people, mostly Europeans, who could be expected to have some familiarity with recent American literature but very little knowledge of American history, especially of early periods, and none of American religion. The Free University of Brussels was, moreover, traditionally and officially hostile toward organized religion.

The essay was written in the summer of 1959, toward the end of "the Eisenhower interlude." On the surface, at least, the Republican boast of "peace, progress, and prosperity" seemed to have some real content. Like most academics, I was partly comfortable and partly uncomfortable in this bland atmosphere. Radical alternatives seemed few, and like others I was trying here to find a "usable past" in the more promising periods of conservative dissent from American optimism. It still seems to me that one could write an interesting history of American nay-sayers, bringing together radicals and conservatives, religious pessimists and the gloomier skeptics.

The lecture was published in the *Revue des Langues Vivantes*, Brussels, XXVI, 3 (May-June 1960), pp. 184-92.

The United States, like some other nations, is very easy to understand—as long as one does not look too closely. To serious students, however, it has seemed a peculiarly baffling and paradoxical country. For a long time, the United States has presented to its keenest observers not one but two faces. One of these has been a rather fat and smiling countenance, the face of a wealthy businessman, cheerful, full of self-confidence and goodwill, perhaps a little simple. The other has been the face of an intellectual, needless to say not a happy face, a face full of the marks of inner struggle and self-doubt.

Since the beginning, the cheerful face has been the easier to see. It has been painted by many Europeans, including liberal Frenchmen in the eighteenth century and radical Englishmen in the early nineteenth. To friendly European observers, the United States in its youth seemed promising and progressive, but somewhat flat. After the rich complexities of European society, America seemed to lack both contrast and nuance. Yet friendly travellers always found much to praise: the high general level of prosperity, the religious toleration, the evidences of material progress, and the exuberant self-confidence. (To less friendly observers, American self-confidence has seemed *more* than sufficient.)

America has been seen as the country of progress, and also of practicality. Our national credo has often been described as a combination of simple moral maxims and easy utilitarianism.

When the philosophy of pragmatism was formulated by Peirce, James, and Dewey in the late nineteenth century, many said that this doctrine, with its activism, its dislike of metaphysical categories, its emphasis on practical results, was *the* American philosophy.

This is the familiar picture. To its admirers, the United States has seemed progressive, optimistic, and practical. Its harsher critics have said almost the same thing in different words: to them American civilization has seemed simple, overconfident, and more than a little dull. The portrait is recognizable; I do not think I have to describe it in detail. This version of American civilization is familiar everywhere, not least in Belgium.

My subject today, as the title announces, is not this tradition of cheerful, practical self-confidence, but the Other American Tradition, the tradition of alienation and self-doubt. We can approach this Other Tradition most easily through literature. At first, it is true, many of our critics and literary rebels have seemed to confirm the standard picture of America. They have agreed with European critics that America is simple, cheerful, and uniform. But they have presented this picture without affection, without even balance or tolerance. They have presented it with extreme distaste, even with dismay. In each of the more vigorous periods of our literature, leading American writers have complained bitterly of American materialism and uniformity, American complacency and intolerance. This complaint has been phrased with every degree of prophetic indignation. Let me present two well-known samples, neither taken from the works of an extremist or exile.

In 1871 Walt Whitman, generally thought of as the bard of expanding democracy, said this:

> I say that our New World democracy, however great a success in . . . materialistic development, is, so far, an almost complete failure in its social aspects, and in really grand religious, moral, literary, and esthetic results.

In 1922 Van Wyck Brooks, then spokesman of the younger critics and now dean of American literary historians, said almost the same thing:

> Considered with reference to its higher manifestations, life itself has been thus far, in modern America, a failure. Of this the failure of our literature is merely emblematic.

One could go on almost endlessly, quoting important American writers who saw with despair the triumph in America of materialism and conformity, the paralysis of art and individuality.

Here, a paradox arises. If it is true that America has been so uniform a country, where have the critics come from? How does it happen that in a country dominated by simple material progress, so many writers have passionately denounced materialism and rejected conventional progress? If America is the country of cheerful conformity, how does it produce, in every generation, so many young men who denounce cheerful conformity?

When one has asked this question, one has begun to look for the Other Tradition. And if one looks at our literature, it is not hard to find.

In 1886 William Dean Howells made a famous statement, a statement which has been much quoted and much denounced. Our novelists, he said, would never deal with subjects like those of the Russian and French novelists. They would concern themselves with "the more smiling aspects of life, which are the more American." This prophecy must have seemed entirely plausible; it has proved to be a colossal error. Actually it is difficult to find a major American writer who deals with the more smiling aspects of life. One has only to mention the names of Poe, Melville, Hawthorne, Henry James, Stephen Crane, Ernest Hemingway, William Faulkner, to demonstrate that whatever else American literature is, it is not a literature devoted to practical progress and external success. The greatest American writers have, in fact, devoted themselves to tragic conflict, and many have specialized in the extremes of inner and outer horror. For many of the most important, the typical method has been not robust realism but symbolism. As Richard Chase and other recent critics have pointed out, one looks in vain for major American analogues of the sturdy, direct English novelists—it is hard to find an American Fielding, or an American Trollope.

Here then, in very simple terms, is the paradox that confronts serious students of the United States. On the one hand it is the country of simplicity, progress, and optimism. Yet it has produced an extraordinary number of gifted men whose works have been characterized by complexity, tragedy, and self-doubt. Many Americans have illuminated one of these pictures; none I think, has explained both at once. Students of literature have devoted a great deal of attention to the alienated intellectual, but they have seldom discussed very seriously the society which he was rejecting. Historians and many social scientists have usually stuck to the story of economic and political success, and doubtless they have been right to do so. Recently, however, it has become conventional even for historians to take some account of literature. This spoils their narrative. In the midst of the chapters on the early nineteenth century, the period of expanding democracy, they must account for Poe, Hawthorne, Melville, and Thoreau. None of these was able to contemplate American society without some sort of serious misgivings. When historians reach the nineteen-twenties, the age of prosperity, the period of Ford and Edison, they must at least mention in passing the fact that American literature was dominated by disaffected writers, some of them self-exiled. If they deal with still another very recent period, the decade after the Second World War, historians reach once more a time of unparalleled American wealth and power. Yet again they have a number of awkward elements to fit into their picture, among them the fact that T. S. Eliot, who detests contemporary civilization, achieved in this period something like official hegemony over American criticism. American literature in all periods refuses to fit into the conventional, progressive, and optimistic synthesis of American civilization.

I am not hoping, in this lecture, to clear up all difficulties and present you with a new synthesis. All I am going to do is to suggest, very briefly, a few of the explanations which have been advanced for the discontent of American literary intellectuals. Perhaps you, looking from the outside at this American problem, may find some interesting insights into our civilization. In the long run, it may well be that the successful synthesis will have to come from a detached European, a new and greater Tocqueville.

The simplest, and one of the most common explanations has been geographical. All the gloom and doubt have come from Europe. American intellectuals have fallen under the spell of European decadence, and they have been seduced away from their natural, American cheerfulness and confidence. This has been seriously implied, not only by defenders of American optimism, but by some of its stoutest enemies. H. L. Mencken, for instance, sometimes seemed to imply that nearly all Americans were cheerful simpletons, and that nearly all Europeans were ripe, sophisticated, aristocratic pessimists.

If this geographical theory were true, if the optimists were the real Americans and the pessimists, doubters, and ironists all Europeans in disguise, some curious results would follow. The English prophets of Victorian progress, Herbert Spencer and his contemporaries with their many followers on both sides of the Atlantic, would have to be essentially American. On the other hand, we would have to count as Europeans not only Henry James and T. S. Eliot but also Ambrose Bierce, H. L. Mencken and that tormented and savage satirist, Mark Twain, not to mention the sad and introspective Abraham Lincoln. Obviously, American doubts and fears are as native as American confidence. Both are American and this means that both have European roots.

A second and equally simple explanation has often been suggested by the fact that much of the self-doubt comes from literary

men. To some defenders of American cheerfulness, this makes it
clear that literary dissent is simply an American phase of the general nineteenth-century counter-attack by the artists against the
bourgeoisie. To put the matter still more simply, discontented
Americans are simply neurotic esthetes. During each of the two
world wars, a few inflamed patriots went even further, suggesting
that literary doubters were almost treasonable, almost insane.
(To many Americans in all periods, it seems obvious that anybody who rejects American civilization *must* be insane.)

It is hardly necessary to say that this last kind of indignation
arises from a lack of imagination and understanding. In their
own way, nearly all of the literary rebels who have denounced
American optimism and uniformity have been intensely patriotic. They have talked about America in the spirit of the prophets
of Israel, lamenting that their country, a chosen people with a
special destiny, seemed at times to be betraying its great mission.
The extreme seriousness and intensity of this kind of American
self-criticism is, I suggest, worth bearing in mind in our search
for origins.

Moreover, the kind of discontent expressed by our writers can
be found well outside strictly literary circles. One must remember to begin with readers as well as writers, not only the author
of *Babbitt,* but the thousands of Babbitts who bought *Babbitt,*
the countless Americans who applaud lectures by their most
damning critics, the thousands who constantly assert the superiority of other countries to their own. Far more than other
countries, Americans, despite their reputation for self-satisfac-
tion, seem to indulge a taste for self-castigation.

If we move for a moment still further from imaginative literature we will find other signs, intangible but suggestive, that
many Americans are uneasy with the progressive and optimistic
synthesis. It is perhaps significant that more and more Americans, when they read history for pleasure, turn away from the
glowing pages of expansion toward the one obviously tragic episode in our history, the Civil War. The favorite popular hero is

not the progressive and optimistic Jefferson, but the brooding, melancholy, and compassionate Lincoln, and it is exactly these elements in Lincoln that are emphasized in popular tradition.

In recent years, many of the sharpest criticisms of American life have come not from poets and novelists but from sociologists, who echo many of the complaints made by literary figures in the past. American life, we are told, is tense and joyless. Our sex habits, our methods of raising children, our amusements betray increasing insecurity. In this age of the organization man, fewer and fewer Americans dare to raise their voices against American conformity. So we are told by books published nearly every week, and it is worth noticing that such books are nearly always favorably reviewed and widely read.

A third explanation of American self-doubt is chronological. American civilization was predominantly cheerful and confident, we are often told, until the twentieth century. After two hundred years of expansion, after a hundred years of freedom from foreign war, we suddenly found ourselves, after 1917, in the midst of a series of tragic dilemmas. Our divisions and passions in recent years, and also the extreme disquiet of some of our intellectuals, are simply a result of shock.

I do not want to deny the truth contained in this argument (so brilliantly stated for instance by Reinhold Niebuhr in his *The Irony of American History*). And yet it is clear that the roots of the Other Tradition lie far deeper than the twentieth century. Our dissenting intellectuals of recent decades have been able to discover useful ancestors. It was the young iconoclasts of the nineteen-twenties who rediscovered Melville, appreciated Henry Adams, and penetrated the comic disguises of Mark Twain. More recently, modern literary intellectuals like Perry Miller have made us realize the force and power of Puritan theology, America's oldest and most systematic theory of human wickedness. Still more recently, some historians have begun to rescue from the easy progressive version of American history some of the wealth of our eighteenth-century political tradition—

active, enterprising, yet profoundly sceptical. The twentieth century and its shocks have indeed caused Americans to reexamine
their own past. When they have done so, they have discovered,
beneath the thick layers of conventional bland optimism, some
of the variety and richness of the Other American Tradition.

Still another explanation for the alienation of American intellectuals developed, naturally, out of Marxism. In the depression
decade, the nineteen-thirties, a good many American writers
turned, usually briefly, toward the Far Left. Here they found
a clear and plausible explanation for some of the discontent
they had long felt. It is important, I think, to remember that a
great many of those who turned toward communism in this
period of economic crisis had already been deeply critical of
American society during the period of economic success.

Now they learned to believe that the apparent successes of
American society were a fraud. Large sections of the people
lived in misery, and the rest were devoted to concealing their
guilt from themselves. American civilization, ruthless and hypocritical, naturally drove sensitive members of the bourgeoisie to
despair.

This explanation is so extremely out of favor today that we
may be in danger of neglecting the fragments of truth it contains. Let us admit immediately that injustices, important injustices, have existed in every period of our history. Yet in comparative terms—and it is hard for historians to agree on any
others—this history is one of the great success stories. The overwhelming majority of Americans have taken for granted the essential triumph of American society. Only certain kinds of people have found life in America, from time to time, hard to accept
or endure. These unhappy individuals, who have written most
of our best literature, appear in periods of prosperity and depression alike. Often they seem to resent our material successes
even more than our failures.

Few of the more impressive spokesmen of literary dissent have
arisen from the socially disinherited groups: the Negroes, the

Southern small farmers, the recent immigrants. More typically, dissenters have appeared in the bosom of the country's conservative families, in the oldest and wealthiest Eastern colleges, in the prosperous farm areas of the older Middle West. The intellectual radicals, the castigators of American conformity, have not always been radicals in economic and social terms, and when they have, their social radicalism has been a secondary element in their general malaise. Conversely, many American radicals—socialists and even communists for example—have dissented only in part from the official purposes of the majority. They have wanted to change the ownership of wealth and the political machinery. Yet some of them, not all, have cordially accepted as their goals increased wealth and also increased uniformity.

In my opinion, the sharpest arrows directed against American complacency, those which have come closest to piercing its tough hide, have not come from the conventional socio-economic left. Social radicalism, on the one hand, and literary or intellectual dissent on the other, both have long and interesting histories in America. The relations of one with the other have been interesting and complex. All I can do now is to insist that they are *not* the same thing, and have not the same sources.

In all these explanations of literary discontent—geographical, chronological, and social, there is some fragment of truth, and still other suggestions have been made which I have not time to review. One more major suggestion, by no means new, seems to me to contain more substance and less error than the rest.

Much of the origin of American intellectual dissent lies, I am convinced, in the history of American religion. In the second and third decades of this century, literary rebels used to trace all the evils they found in American life—conformity, bigotry, compulsive hard work, joylessness—to one simple source: Puritanism. These rebels knew very little about Puritanism, and yet they were partly right. If they had known a little more about Puritanism, or about Protestantism in general, they would have come, I think, still closer to the truth. They would have seen that they themselves were still more the heirs of the Protestant spirit than was the society they were attacking.

If one reads the literature of American intellectual dissent in any period, one encounters a set of characteristics which can hardly fail to remind one of the seventeenth-century intellectuals of Massachusetts Bay, and of their direct heirs in the history of American religious innovation from Jonathan Edwards on. These are the suggestive traits: a habit of agonized self-doubt, a deep suspicion of material appearances, a positive hatred of blandness and complacency, and above all a most intense and even painful seriousness about oneself, one's country, and its mission.

These traits seem to me to be obviously rooted in the Protestant past. This is no less the case when one finds the telltale characteristics outside the ranks of preachers and philosophers, among poets and novelists, critics and social scientists, many of them agnostics or even militant atheists. Articulate beliefs change rapidly, in years of intellectual revolution. Emotional tendencies, transmitted from generation to generation, endure much longer.

When I speak about transmission from generation to generation, I do not mean to subscribe to any old-fashioned theory of the inheritance of ideas. I am not suggesting that Protestant seriousness and intensity were "in the blood" of American rebels. They were transmitted through less mysterious channels.

Throughout most of our history, we are at last beginning to remember, most Americans have derived their general ideas from *two* main sources, politics and religion. In their newspapers and in political speeches they have been told about the success of American institutions and the Utopian future of American society. In their churches every Sunday, however, and in countless tracts and incessantly repeated hymns, they have heard a different story. Some of them used to believe that man was totally depraved. With the decline of Calvinism and the rise of evangelical Protestantism, more learned to believe that every man had a chance for salvation. Yet man remained, for very many Americans, a poor sinful creature whose only hope came from Divine mercy. And however inadequate the individual might be to bear his burdens, duty lay heavy upon him. One has only to dip into American memoirs of the nineteenth century to realize how many adolescents, even this late, struggled with themselves to try to find some faint trace of spiritual worthiness. For many, this sort of inner experience was the most important thing in the world; a close second was the external triumph of American secular and religious institutions. Moreover these two concerns were not as separate as they might seem: when American society, swollen with success, turned towards materialism it was in danger of losing its soul. This, though they have not always used this language, has been the principal message of the most effective American dissenters in all periods.

In calling attention to the consistent pattern of American literary discontent, I do not wish to minimize the importance of the cheerful and confident outlook of the American majority. This is the major tradition, and the larger historical force. Yet the Other Tradition, the tradition of agonized self-doubt, is equally real. Each of these facets of our intellectual history is necessary to explain the other. One cannot fully grasp the power of the American majority unless one sees it through the eyes of the dissenters, and one certainly cannot understand the dissenters unless one knows something of the society they are criticizing. To understand America in any period, one must study such pairs of opposites as Benjamin Franklin and Jonathan Edwards, Andrew Jackson and Nathaniel Hawthorne, Theodore Roosevelt and Henry James, even Dwight Eisenhower and William Faulkner.

So great is the power and success of the major, self-confident American tradition that it has often seemed as if the Other Tradition might disappear. It may seem so, to some of you, now. It seemed so recently to some people in America. Yet just when things began, in the last few years, to seem almost unbearably cheerful, the Russians came to the rescue. The first sputniks produced in America one of those recurrent waves of national self-searching that seem necessary for our balance. Today no statement is more conventional, in many quarters in America, than the flat assertion that American education is a costly failure which must be completely and drastically revised. Such complaints may be exaggerated, but they are bracing, and they are part of a great tradition.

For those who do not like to have their skies entirely blue, there is always hope; new storms are always forming on the horizon; there are some there now, quite large and quite real. But the exponents of the Other Tradition do not need external storm-clouds. For them the biggest tempests arise within. However great our successes, however peaceful the climate—and let us hope it remains as peaceful as it looks—American confidence and American self-doubt will continue, I believe, their exhausting but creative battle. Without this battle, our history would be very different, and our literature would hardly exist.

· 4 ·

The Recovery of
American Religious History

This essay grew slowly, as did my interest both in religion and in American religious history. In 1949 I published my first book, *Protestant Churches and Industrial America*. This book, first written as a thesis and strongly influenced by my thesis director, A. M. Schlesinger, Sr., looked at American religious history from a neutral and thoroughly secular point of view, emphasizing the effect of religion on society rather than religion's inner content. However, the work I had done with Perry Miller had suggested a quite different view, emphasizing theological content, looked at with deep sympathy by a man who considered himself an atheist. From the late 1950s on I was one of those described in the essay as "interested in religion." In 1959 I was invited to take part in a symposium of Protestant seminary historians of American religion, and was shocked to find some of them denouncing "secularist" religious history and insisting that religious history must be seen as chronicling the work of the Holy Spirit. From then on I tried to work out a position between the rigidly secularist and churchly camps, looking at religion and religious history with sympathy somewhat short of clear commitment. In 1962 I read a version of this essay to a meeting of the American Studies Association at Fresno State College.

In 1964 I rewrote the essay and submitted it to the *American Historical Review*. The very mixed reaction of the editors and of the outside experts to whom they submitted the essay was reported to me with admirable frankness and copious excerpts (though no names) by the editor, Professor Stull Holt. It was clear that the article had affronted and upset several critics, who completely denied the existence either of the contemporary revival of religion or of the recovery of religious history. On the other hand one of the outside critics said that the article was not clearly theological enough. One commentator advised publication because the essay would "stir up the animals."

This last reader proved wrong. When Professor Holt, after long hesitation decided—I think with some courage—to publish the essay despite all the criticism, I received no negative and many positive comments. I believe it has been cited and quoted more than any other of mine.

Since I wrote this essay many of the circumstances that I had seen as contributing to the recovery of religious history have sharply changed. The popular revival of religion waned, then leveled off. The mainstream churches lost out on either side, to the cults of the sixties and to older evangelical and fundamentalist movements. Neo-orthodoxy fell out of favor and perhaps began a comeback. Intellectual history, especially of the analytic kind mentioned in the article, lost ground to "the new social history."

Yet the growth in quality and quantity of American religious history seems to me to have continued, and many of the particular gaps to which this article pointed have been filled. Sydney Ahlstrom provided the missing synthesis, and Catholic historians, sometimes citing this article with approval, showed a new vitality. Perhaps the principal recent tendencies have been to look less at the intellectual content of religion and more at its social history, and to shift attention from the once-dominant churches of the Protestant middle class to the immense variety of American popular religion. It is my impression that regardless of their religious or non-religious standpoints, American historians today are much less likely than they were in the sixties to question the importance of religion for understanding the American past, and that this is especially true among young scholars.

For the study and understanding of American culture, the recovery of American religious history may well be the most important achievement of the last thirty years. A vast and crucial area of American experience has been rescued from neglect and misunderstanding. Puritanism, Edwardsian Calvinism, revivalism, liberalism, modernism, and the social gospel have all been brought down out of the attic and put back in the historical front parlor. Out of monographic research on these and other topics, it begins to be possible to buld a convincing synthesis, a synthesis independent of political history, though never unrelated to it.[1]

Even for those students of American culture who do not find religious thought and practice intrinsically interesting, knowledge of religious history has become a necessity. This is most obviously the case for those interested in American intellectual history. In the first place, the recovery of American religious history has restored a knowledge of the mode, even the language, in which most Americans, during most of American history, did their thinking about human nature and destiny. In the second place, the recovery has necessitated, though it has not yet really affected, a reorganization. Obviously the categories of V. L. Parrington, once so satisfactory, will no longer work. One cannot, for instance, oppose "French" liberalism to Calvinist conservatism as the poles between which to classify both political and religious thought in the early national period. What is one to do with orthodox clergy who supported the American and for long defended the French Revolution, with Whig conservatives who were Unitarians, or with doctrinally conservative Presbyterians who took the side of Jackson in politics? There are too many exceptions: they destroy instead of proving the rules. Nor can one talk any longer, without important qualifications, about an "American faith" in which optimism about man is inescapably linked to democracy. To insist on such a link, one has to rule out of the American democratic tradition not only such "belletristic" aberrants as Henry James, Hawthorne, Poe, and Melville, not only such political exceptions as Calhoun or Henry Adams, but also John Adams and Madison, which is difficult, and both Lincoln and Mark Twain, which is downright impossible.[2] To summarize the central American tradition has become a far more difficult task than it once was, and a far more interesting one.

Restoring a language and shaking up a set of categories are not the only changes wrought in intellectual history by the recovery of religious history. By analogy the work of religious historians illuminates two major perennial problems of the American intellectual historian. The first of these is the relation of American to European thought. Obviously American church history cannot be studied without reference to the Reformation, and thus to European thought since (or perhaps before) the patristic period. Yet, as Tocqueville, Schaff, and Bryce saw and as lesser European commentators have often not understood, American religion cannot be forced into European categories. Like many other kinds of American experience, religious experience serves both as link and barrier between the continents.

The other problem of intellectual history illuminated by the example of religious history is the even more difficult one of the relation between ideas and institutions. For this the history of American Protestantism, with its long effort to institutionalize successive religious impulses, offers also some highly interesting suggestions.[3]

That part of literary history which lies closest to intellectual history has been transformed with it, or even before it.[4] At an opposite pole in American studies, the analysis of American class structure has been enriched. Sociologists must study church history and even theology. Simple lines between denominations will no longer do; to locate someone in American society it is necessary to say what *kind* of a Baptist or Presbyterian he is, and where, in religious and other terms, he comes from.[5] Historians of our two greatest political crises have revived a religious interpretation of each.[6] Theorists of American foreign policy—including some theorists not far removed from the scene of action—frequently invoke kinds of thought that were originally theological.[7]

Of the several meanings intended by the title of this article most historians will, I think, readily admit the fact of an increased emphasis on religious history. Many have also observed—whether or not they have approved—the emergence of a more sympathetic assessment of American religious experience. Here consensus stops; historians disagree about the causes of these related changes. Part of the disagreement is inevitably ideological; part arises from the complexity of the subject. In American historiography, as in American religion, categories shift and change. Yet categories are necessary, and a look back at major names and dates suggests a few.

The recovery of American religious history really began in the 1930s. In the twenties, nineteenth-century "scientific" history was being challenged by the brilliant agnostic relativism of Becker, the fervent progressivism of Parrington, and the somewhat selective determinism of Beard.[8] "Puritanism," and the larger religious tradition loosely associated with it, was under heavy attack inside and outside historical circles. Harold Stearns explained in 1922 that there was no article on religion in his *Civilization in the United States* because he could find no one interested in the topic.[9] This was a Menckenesque exaggeration, but it was true that American religion, aside from the dramatic forays of the fundamentalists, did not look very interesting. The dominant liberal Protestantism was reaching the end of a long, igno-minious, and unsuccessful effort to accommodate its teaching at any cost to the ultrasecular culture of the day.[10]

Seminary historians played their parts in this effort at accom-modation, trying hard to follow the lead of the dominant uni-versity historians. Most of them, attempting to be neutral and "scientific," produced factual monographs limited by denomina-tional lines. The two best-known general histories of American religion written during the period, those by H. K. Rowe (1924) and W. W. Sweet (1930), followed the lead of standard secular interpretation. Rowe emphasized the growth of liberalism and religious freedom, Sweet, the frontier. Both works were respect-able; neither was highly original.[11]

In the thirties, when the recovery began, a student who wanted a treatment of American religious history with some feeling for theology had to go back beyond the twenties to such books as F. H. Foster's *Genetic History of New England Theology* (Chi-cago, 1907) or Leonard Bacon's *History of American Christianity* (New York, 1897). If he wanted to investigate religious experi-ence, he invariably started with William James's unique and curious classic of 1902. Only for the topic of religion and social class—a topic that interested him greatly—did he have a first-rate recent work, H. Richard Niebuhr's *Social Sources of Denomina-tionalism* (New York, 1929), which applied to American religion the insights of Max Weber and Ernst Troeltsch.

In this atmosphere, the recovery of American religious history was begun by the only people in a position to undertake it, the immensely energetic secular scholars of the day. To men schooled in objective examination of, as nearly as possible, *all* the data, religion was too big to be ignored in the flippant manner of a Harold Stearns. Herbert Schneider, who occupied a chair of religion at Columbia University, treated the American religious past with much learning.[12] At the end of the period Ralph Henry Gabriel in *The Course of American Democratic Thought* (New York, 1940) gave religion a much more active constituent role in intellectual history than had Parrington, though he too discussed religious ideas without much theological analysis.[13]

The best-informed and most influential student of American social history was A. M. Schlesinger, who says with great candor in his recent autobiography that the central questions of religious thought have never held much interest for him.[14] Accepting nonetheless readily the importance of religion for most Americans in the past, Schlesinger directed toward this field the efforts of many students, all of whom did their best to penetrate its obscure shadows with the clear light of objective research. In an influential essay of 1932,[15] Schlesinger himself applied to religious history the insight that was shortly to inform his *Rise of the City*. In the neglected period of the late nineteenth century, he said, American religion had undergone a series of highly important reactions: to Darwinism, higher criticism, comparative religion, and, most important of all, to the challenge of the city.

Commenting much later on this essay and its influence, one of the current group of able seminary historians finds its insight useful and important even from his own very different point of view.[16] Nevertheless, says this later critic, Schlesinger's description of the church transforming itself in response to the urban challenge, like Sweet's description of the church reacting to the frontier, makes the role of the church too passive and neglects inner changes not entirely determined by these external pressures. Whatever the merits of this criticism and the now widespread view it implies, the rescue of religious history was largely begun, as it had to be, from a secular point of view not unlike that of Schlesinger.

Though secular and academic historians dominated this stage of the recovery until after World War II, two quite different tendencies of the thirties foreshadowed a challenge to this domination. The first was the expansion and reorientation of the study of American literature. Still full of the revolts and rejections of the twenties, but discontented with the simple categories of the past and only partly satisfied with the Marxist stereotypes of the present, many of the best young students were fascinated by the complexities, doubts, and inner struggles of writers like Melville, Hawthorne, and James. Nothing, they found, could be farther from the truth than the facile dictum of Howells, still faithfully echoed in very recent years, that American literature dealt characteristically with the surface, "smiling aspects of life." Sometimes venturing beyond American literature into one of the new programs in American studies, students coming from literature departments encountered (more often than history graduate students of this period) the infinitely complex world of recent historical thought, from Mannheim to Whitehead or Collingwood. Admiring complexity and un-compromising intellectual struggle, some of them discovered a new field: theology. To literary intellectuals of the thirties, the-ology was approachable partly because it seemed to have so little to do with religion, especially the religion of the First Methodist Church in the generic home town.

In the study of religious thought in American literature or culture, students of this kind found gifted mentors. One was F. O. Matthiessen, immensely attractive as a scholar and leader to this generation, and far more passionately interested than most of his students in the relation between social radicalism and religious commitment.[17] Even more important for the sys-tematic study of American religious thought was Perry Miller. In 1928, consciously defying the advice of his own teachers and the Menckenian prejudices of the times, Miller had begun his gigantic excavation of Puritanism.[18] In many ways a product of the alienated and tormented twenties, an atheist and something of a radical, Miller yet went to work to rescue Puritan and Calvinist thought with a relish for all its paradoxes and tensions, and with a zeal, sometimes with a polemic intensity, comparable almost to that of Edwards himself. Surely the result of his labor, deepening the history of the American mind in a chronological sense as well as in others, must be one of the most enduring as well as one of the strangest monuments of the radical thirties.

The other development that began in the 1930s to suggest the rise of a new kind of religious history was the turn toward neo-orthodoxy within Protestantism itself. Like Edwards and many other American religious figures, Reinhold Niebuhr, the central American figure in this diverse movement, drew heavily on contemporary European thought. But, again like Edwards and many other prophets, he started by reacting to the smug society he encountered around him. It was not Auschwitz or Hiroshima, but Detroit in the twenties that started Niebuhr on the road away from accommodation and optimism and toward a belief in a world under judgment.[19] Thus there is something in common between Niebuhr's rejections and those of contemporary literary critics, though very little that is common in their respective affirmations.

Two very different works in American religious history reflected a neo-orthodox emphasis in this period of beginnings. The first was Joseph Haroutunian's study of American Calvinist theology, *Piety vs. Moralism* (New York, 1932). Only a book written, like this one, from a neo-orthodox point of view could at this moment have restored meaning to the long-neglected family fights of New England divines, distinguishing in their thought between the new and the merely orthodox. The other historical work that reflected the new theological tendency was H. Richard Niebuhr's *The Kingdom of God in America,* eventually to become one of the most influential books in the whole field. In his introduction, Niebuhr criticized his own much-admired *Social Sources of Denominationalism.* A sociological approach like that of the earlier book, he now said, "helped to explain why the religious stream flowed in these particular channels," but failed to "account for the force of the stream itself."[20] American Christianity should be treated not as a series of institutions but as a prophetic movement, never completely embodied in any institutional forms, liable to decay but capable of perennial self-renewal. This deeply Protestant view of church history was to influence many of the ablest religious historians of the next period. Shorn of some of its religious meaning, Niebuhr's suggestion was usable by historians of other kinds of ideas and institutions.[21] Might not his idea of a cycle of reform, organization, decline, and renewal illuminate the fate of many kinds of ideas in a fluid and energetic society? One might suggest progressive education, temperance, conservation—almost any American movement, perhaps including political democracy itself—as test cases.

In the period since the Second World War, the period of fruition that succeeded this one of preparation, all the influences already active continued to operate. Social historians, among them students of both Schlesingers, of Oscar Handlin, of Richard Hofstadter, and others, continued to deal with the history of American religion from a largely secular point of view. A flood of monographs continued to analyze American literature in more and more specifically religious terms. The influence of neo-orthodoxy, spreading like other major movements in American thought from a small center ever more widely, affected historical writing on all sorts of subjects. The clear and acknowledged influence of Reinhold Niebuhr on Arthur Schlesinger, Jr., C. Vann Woodward, and George F. Kennan suggests the dimensions of this periphery.[22]

Three new influences must be added to the list. The first is the development of a new kind of intellectual history, or, more accurately, the revival of an old one. The new intellectual history places more emphasis on the analysis of ideas and less on description of their social antecedents.[23] Needed and overdue, this tendency may sometimes have gone too far, detaching one part of human experience from another in a somewhat mechanical manner, and talking too simply about the influence of one book on another. In any case, much of the new intellectual history has avoided the opposite error of treating ideas, religious ideas included, as simple responses to clearly identifiable stimuli.

The second postwar development that affected religious history came from an opposite quarter and was perhaps complementary. Sociologists and social historians, among them David Riesman and Oscar Handlin, developed a new kind of analysis and criticism of American society, emphasizing the search for identity among the pressures of a plural, yet sometimes conformist society. To some students of American religion, this suggested a new interpretation of the past and present role of religious groups.[24]

The third new influence, pervasive and complex, was the religious revival of the 1950s. The nature and even the existence of this revival have been endlessly debated. Was there any connection between such phenomena as swelling church statistics, highly successful traditional revivalism, best-selling and sugary "peace-of-mind" manuals, semi-official association of God with American foreign policy, and gingerly, reluctant inquiry into the religious turn of "intellectuals" carried out by the editors of the *Partisan Review?* Could any of these have any connection with the devastating disjunctions of Karl Barth or the tragic view of history propounded by Reinhold Niebuhr? Was this really a revival of religion, or only a search for identity on the part of third generation immigrants or other-directed exurbanites?

At least three aspects of this complex phenomenon must be taken into consideration for our present purposes. First was the new realization of American religion's numerical growth, both short-term and long-term. According to widely cited reports, more than 60 per cent of the population were now church members, as opposed to 5 per cent in 1776 and 35 per cent in 1900. How far to accept either the accuracy or the implications of these figures was a complex question. Yet it was clear that one could no longer talk about American religion as something that used to be important. At least according to the most concrete indexes—numbers, buildings, and money—it was a spectacular success. One exaggerated but suggestive interpretation said that rapid growth had from the beginning determined the whole nature of American Protestantism. The American churches were not branches of European Christendom, but new churches, with the good and bad characteristics of new churches everywhere.[25]

A second element of the revival was the continuing vitalization of theology. One historian had this to say:

> One must go back to the sixteenth century to find an era of equal theological fertility and creativeness. In America it is at least a century and a half since theologians held a position of such importance in our national thought. Now that John Dewey is dead there is in the United States scarcely a single philosopher of public eminence who is confronting the traditional "problems of man" as comprehensively as are at least a half-dozen theologians.[26]

Often the theological renaissance and the popular growth seemed opposed rather than complementary; no one criticized so harshly the easy, amorphous popular "faith in faith" as those who had for some time been demanding faith in something more specific and difficult. Yet some highly sophisticated historians of religion concluded that this revival, with its depths and shallows, its center and periphery, was not altogether different from revivals in the past.[27]

A third fact about this revival, which did seem to differentiate it somewhat from its predecessors, was the complexity of its effect on American intellectuals. As with some of the earlier revivals, a great many intellectuals remained hostile to this one in all its aspects. Others, including poets, novelists, and a few historians, stood fundamentally within it. A large number, however, and the most important group for our present purposes, belonged in neither of these opposing camps. As Sydney Ahlstrom put it, "curiosity about religion" penetrated intellectual circles far beyond the ranks of the committed. This "curiosity" might "lead a person merely to taste some popular book, to take up the study of some religious poet, or to dedicate himself to a lifetime vocation of religious research."[28] In the middle of the twentieth century, that is, not everyone could find himself at home either among believers or militant secularists.

The consequences of these diverse changes were themselves diverse. The first was a new understanding of the nature of the mainstream of American religious history. In the thirties, many students of American religion had understandably admired Puritanism and Edwardsian Calvinism and disliked revivalism. Now it seemed clear that (as Miller's own work had indicated) both Covenant Theology and Consistent Calvinism had been brilliant, heroic, but unsuccessful attempts to channel the turbulent flood of American religious energy. The mainstream, for better or worse, had been revivalistic, emotional, even somewhat pragmatic. Faced with the task of evangelizing an unchurched continent, of combatting not only infidelity but barbarism, first on the frontier and then in the cities, the American churches had indeed compromised theological differences. So far had this process gone that they had almost lost their distinctive task and message. But when adaptation had gone too far, self-criticism had restored some balance, in the early eighteenth century, in the late nineteenth century, again in the mid-twentieth century.

Obviously this reorientation owed much to the past insights of Sweet, Schlesinger, and Richard Niebuhr. It owed much also to the new investigation of revivalism by such lay historians as W. G. McLoughlin, Jr., and Bernard Weisberger.[29] Still more important, however, were the insights of Timothy Smith, himself a minister, and the seminary historians Sidney Mead and Winthrop S. Hudson. All these saw the past tasks of American religion as men who had a more than academic concern with its present pastoral duties. And all looked at revivals from the midst of a period of revival.

Smith found much of the vitality of American nineteenth-

century religion in the perfectionist tradition. It was the belief
in the possibility of perfect holiness, he argued, that furnished
the energy for many reform crusades. Somewhat more complex
in their loyalties, Mead and Hudson emphasized both the failures
and successes of the revivalist tradition, failures and successes
inseparable from those of American culture.[30]

The second consequence of the new religious surge owed more
to the theological renaissance than to the popular increase. This
was the demand on the part of a number of the seminary historians for a new kind of church history, emancipated from the
long subservience to "positivist lay historians." With varying
degrees of fervor, a number of manifestoes including more than
one presidential address to the American Society of Church History called for a separate "church history." This must be the
history of the church eternal, invisible, and universal; it must
indeed be a narration of the continuing work of the Holy Spirit
on earth.[31]

To the more extreme of these statements, lay historians and
even some of the more moderate "church" historians themselves
could raise several objections. In the first place, it seemed strange
for Protestants to be quite so confident about the exact operation
of the Holy Spirit or its limits. Was not church history, by some
of the definitions now suggested, history itself? If so, could one be
quite certain to whom it was given to understand it?[32] Second, on
quite a different level, the demand for a separation from "positivist lay historians" seemed sometimes to ignore the fact that historians so described had done much of the research on which any
interpretation of the American religious past, from any point of
view, had to depend.[33] Third, this same demand for sharp separation seemed to ignore the nature of the current revival. Despite
the great either-or's of some of its theologians, its effect had been
to blur, not to sharpen the line between believers and non-
believers, among historians as elsewhere. To draw a line between
believing "church historians" and "positivist lay historians" had
become impossible. Examples of both could be found, but one

could also point to historians who combined impeccable secular academic credentials with seminary training, ministerial experience, or explicit religious affiliation. Many others had been touched to one degree or another by the revival of "interest in religion," and some who had not could hardly be called "positivists." The nature of the current religious situation had made religious classification of historians impossible. No one could say with precision where, in religious terms, the best new writing was comng from.

It is a little easier to say where the best work in religious history was *not* coming from. The groups that recently seem to have contributed least are opposites: atheists and Roman Catholics. No one in the recent period has examined American religion with the scholarly love-hatred of H. L. Mencken at his best. An explanation is suggested by Martin Marty's sketch of the history of the American infidel.[34] By the early twentieth century the commitments of American Protestantism had become so amorphous that there was nothing left to hate. Perhaps one of the tests of the depth of neo-orthodoxy is the question whether it will produce a neo-atheism.

Since the 1955 bombshell of Monsignor John Tracy Ellis, Catholic intellectuals have been discussing the failure of American Catholicism to participate proportionally in American intellectual life in general. Some of them have also suggested that American Catholic history in particular has been slighted, and especially the history of the recent period of rapid Catholic increase.[35] Undoubtedly the social explanations suggested by Ellis, arising from the immigrant past of the American church, contain part of the answer. But some other suggestions have more relevance for our present purpose. One Catholic writer blames Catholic "formalism," "the tendency to see the world as 'finished' and all things in it as obvious in their essence and meaning," and also Catholic "Authoritarianism," of which one result is "the illusion of a neat universe in which nothing eludes the conceptions of a searching mind."[36] Still another suggests that the failure arises partly from the great difficulty, for American Catholics, especially since the crisis ending with Leo XIII's letter on Americanism, of carrying on a searching dialogue with non-Catholic American culture.[37]

These descriptions of Catholic difficulties seem relevant to some of the conclusions of this article. A point of view that has proved extremely fruitful both for Protestant and for nonreligious American historians (and for many who lie, as we have seen, between these two categories) is that suggested by Richard Niebuhr. From this point of view, religious impulses are never fully embodied in religious institutions, and the unity to be found in American church history must be found in a cycle of renewal and decline. Obviously, it would be impossible for a Catholic historian of Catholicism to take exactly this point of view. It may be that in this period of new openings, an analogous point of view may be found from which American Catholics can look freshly at their own church, and at American religious history in general.

A partly relevant parallel is offered by American Jewish history. Though historians of American Judaism, like their Catholic colleagues, have bewailed the slowness of development in their field, the problem seems almost opposite.[38] Much of the American Jewish history that has appeared in answer to such complaints deals successfully—not without internal friction and dispute—with exactly the problem so difficult for Catholics: the adjustment of an old and international religion to a Protestant or post-Protestant national culture. Where historians of American Judaism have been less successful, according to some critics, is in dealing with belief and doctrine. This may well arise from the comparatively nondoctrinal character of Judaism itself. Again, perhaps some variant of the approach that has proved fruitful for Protestantism may further illuminate the history of Judaism in America. If so, it will demand both an understanding of the religious stream and a knowledge of its secular channels.

The recovery of American religious history has been the work of thoroughly secular academic historians and also of believers, so far usually believers in some kind of Protestantism. The recent revival of religion has restored something of a balance between these two groups, and thereby it has greatly benefited American religious history. It has done this by restoring depth and variety rather than dogmatism. In America at least, most good history, whether of religion or anything else, has been written by people who are respectful of data, imaginative in dealing with many kinds of experience, and open to new insights—even incomplete and shifting insights. History written by those who confidently describe a single grand design, whether providential, evolutionary, or economic, sometimes impresses, but seems not to endure. Many, though not all, of the best recent historians of American religion do indeed believe that there is purpose in history. Of those who do believe this, few if any suppose that they understand this purpose in any detail. In dealing with the religious past, it is not ordinarily those "interested in religion" who sound dogmatic and defensive today, but rather the more rigid of the behaviorists, Freudians, and economic determinists.

Religious history, in any of the possible meanings of this term, is by no means sweeping all before it, any more than is religion itself. Rather, the revival of both has brought American history back into the great dialogue between secular and religious thought. It is to this dialogue, after all, that American culture itself owes much of its vigor and complexity.

Notes

1. Such a synthesis is not yet published. Most of the important contributions to the recovery have as yet been monographic or critical. Students have, however, been provided with a superb bibliography, a first-rate atlas, and one of the most illuminating of source collections: Nelson R. Burr, *Critical Bibliography of Religion in America* (2 vols., Princeton, N.J., 1957); Edwin S. Gaustad, *Historical Atlas of Religion in America* (New York, 1962); H. Shelton Smith *et al.*, *American Christianity: An Historical Interpretation with Represen-*

tative Documents (2 vols., New York, 1960). The Burr bibliography makes it unnecessary, as it would in any case be impossible, to mention all the significant works in any category discussed in this article.

2. Henry Nash Smith, *Mark Twain, The Development of a Writer* (Cambridge, Mass., 1962), seems to have the last word on the much-argued question of Clemens's pessimism. A recent contribution to the large and growing literature on Lincoln's complex religious views is William J. Wolf, *The Almost Chosen People* (New York, 1959). Though the mature views of the two men are in many ways opposite, I think it legitimate to call both post-Calvinist.

3. See below, pages 73-74.

4. See below, pages 84-85, and also the section, "Religion and Literature," in Burr, *Critical Bibliography,* II, 847-953.

5. A good short list of works on "Church and Class" by sociologists and historians will be found *ibid.,* 606-10. An example of effective use, by a historian, of religious categories for social analysis is Lee Benson, *The Concept of Jacksonian Democracy: New York as a Test Case* (Princeton, N.J., 1961), esp. 186-207.

6. A religious interpretation of the Revolution is reasserted by Carl Bridenbaugh, *Mitre and Sceptre: Transatlantic Faiths, Ideas, Personalities, and Politics, 1689-1775* (New York, 1962). A religious, or partly religious explanation of the Civil War seems to me to rest on two assertions: that serious and intractable moral conflicts were important in causing the war and that in nineteenth-century America such conflicts were peculiarly difficult to avoid or compromise because of the dominance of evangelical Protestantism in both sections. The importance of the moral conflict is implied by much though not all recent writing on slavery and antislavery, and directly argued in the well-known articles of 1946 and 1949 by Bernard DeVoto and A. M. Schlesinger, Jr. (Bernard DeVoto, "The Easy Chair," *Harper's,* CXCII [Feb., Mar. 1946], 123-26, 234-37; A. M. Schlesinger, Jr., "The Causes of the Civil War: A Note on Historical Sentimentalism," *Partisan Review,* XVI [Sept. 1949], 969-81.) The distinct importance of religion in sharpening the conflict is forcefully argued, with respect to the South, by Charles G. Sellers, Jr., "The Travail of Slavery," in *The Southerner as American,* ed. *id.* (Chapel Hill, N.C., 1960), 40-71.

7. McGeorge Bundy calls Reinhold Niebuhr "probably the most influential single mind in the development of American attitudes which combine moral purpose with a sense of political reality," though he is not uncritical of Niebuhr. (McGeorge Bundy, "Foreign Policy:

From Innocence to Engagement," in *Paths of American Thought,* ed. Arthur M. Schlesinger, Jr., and Morton White [Boston, 1963], 293-308.)

8. In an interesting article tracing schools of American intellectual history, Robert Alan Skotheim suggests that the school of historians dominant in the twenties and afterward tended to regard some ideas as determined by the socio-economic environment, and others as possessing autonomous causal importance. Religious ideas, to which men of this school were generally unsympathetic, were usually in the first of these categories, while scientific ideas and proposals for social reform tended to be placed in the second. On the other hand, says Skotheim, some later writers including Perry Miller seemed to make religious thought autonomous and causative, and to treat opposing secular currents as environmental in origin. (Robert Alan Skotheim, "The Writing of American Histories of Ideas: Two Traditions in the xxth Century," *Journal of the History of Ideas,* XXV [Apr.-June 1964], 257-78.)

9. *Civilization in the United States,* ed. Harold Stearns (New York, 1922), v-vi.

10. For excellent accounts of American religion in this period, see Robert T. Handy, "The American Religious Depression, 1925-1935," *Church History,* XXIX (Mar. 1960), 3-16; and Winthrop S. Hudson, *The Great Tradition of the American Churches* (New York, 1953), 195-225.

11. The foregoing paragraph and some other parts of this article owe much to the illuminating essay on "Church History" by George Huntston Williams in *Protestant Thought in the Twentieth Century,* ed. Arnold S. Nash (New York, 1941), 147-78.

12. Herbert Schneider, *The Puritan Mind* (New York, 1930). In his later *History of American Philosophy* (New York, 1946), Schneider himself generously criticizes this early work in the light of Perry Miller's later research. (See pages 28, 29.)

13. Skotheim points out that Gabriel treated American religion with great respect even in essays published in the mid-twenties. Gabriel's interpretation of American intellectual history seems to me to belong neither to the dominant secular and environmentalist movement, nor to the later countermovement to which Skotheim assigns it, but to have some characteristics of both. (Skotheim, "Writing of American Histories of Ideas," 275-77.)

14. A. M. Schlesinger, *In Retrospect: The History of a Historian* (New York, 1963), 193.

15. *Id.,* "A Critical Period in American Religion, 1875-1900," *Proceedings of the Massachusetts Historical Society,* LXIV (1930-32), 523-47.
16. Robert T. Handy, "The Protestant Quest for a Christian America, 1830-1930," *Church History,* XXII (March 1953), 10.
17. See *F. O. Matthiessen 1902-1950: A Collective Portrait,* ed. Paul M. Sweezy and Leo Huberman (New York, 1950).
18. See Miller's introduction to the paperback edition of his *Orthodoxy in Massachusetts* (Boston, 1959), xvii. Other scholars, some of them preceding Miller, played some part in the reassessment both of the Puritans and of Edwards, but I believe few would deny him the major role in this enterprise.
19. See Reinhold Niebuhr, *Leaves from the Notebook of a Tamed Cynic* (New York, 1930); June Bingham, *Courage to Change, An Introduction to the Life and Thought of Reinhold Niebuhr* (New York, 1961), 129-39.
20. H. Richard Niebuhr, *The Kingdom of God in America* (New York, 1937), vii.
21. This is suggested by John Higham, "American Intellectual History: A Critical Appraisal," *American Quarterly,* XIII (No. 2, 1961), 232.
22. See Schlesinger, "Causes of the Civil War"; C. Vann Woodward, "The Irony of Southern History," *Journal of Southern History,* XIX (Feb. 1953), 3-19; George F. Kennan, *Russia and the West under Lenin and Stalin* (Boston, 1961), *passim.* The more direct influence of neo-orthodoxy on the writing of European history, especially religious history, is excellently analyzed by E. Harris Harbison, "The 'Meaning of History' and the Writing of History," *Church History,* XXI (June 1952), 197-207.
23. Two articles by John Higham reflect the development discussed here. In "Intellectual History and Its Neighbors," *Journal of the History of Ideas,* XV (June 1954), 339-47, he distinguishes between "internal" and "external" intellectual history in a neutral manner. In "American Intellectual History: A Critical Appraisal," he seems to come down on the side of more internal analysis and specifically relates this tendency to the recent rise of American religious history. The same tendency is discussed in detail by Skotheim, who finds that by 1950 both the older and the newer historians were turning away from environmentalism and toward a somewhat more autonomous treatment of ideas. Skotheim attributes this change in large part to the failure of relativism to prove adequate in the political crisis of the 1940s. This analysis seems to me sound except that the change referred to was under way in some quarters before

that crisis. (Skothcim, "Writing of American Histories of Ideas," 277-78.)

24. See, for instance, Will Herberg's acute and influential *Protestant-Catholic-Jew* (Garden City, N.Y., 1955).

25. Franklin H. Littell, *From State Church to Pluralism* (New York, 1962). For the percentage figures, see *Yearbook of the American Churches*, 1963 ed. (New York, 1963), 276, and 1933 ed. (New York, 1933), 99. As the editors of these compilations point out, criteria of membership vary drastically from church to church and period to period. Probably the statistics of the recent growth are more acceptable than the older ones. S. M. Lipset argues plausibly that American religion has experienced a "continuous boom" from the beginning rather than a specially sharp recent increase. (S. M. Lipset, *The First New Nation* [New York, 1963], 144-47.)

26. Sydney Ahlstrom, "The Levels of Religious Revival," *Confluence*, IV (Apr. 1955), 41.

27. A generally hostile assessment of the popular revival can be found in Martin Marty, *The New Shape of American Religion* (New York, 1959), and a still harsher one is A. Roy Eckhardt, *The Surge of Piety in America* (New York, 1958). A more balanced treatment, in my opinion, is Ahlstrom's "Levels of Religious Revival." The revival of the fifties is related to earlier revivals in Timothy L. Smith, "Historic Waves of Religious Interest in America," *Annals of the American Academy of Political and Social Science*, CCCXXXII (Nov. 1960), 9-19.

28. Sydney Ahlstrom, "Theology and the Present-Day Revival," *ibid.*, 27.

29. William G. McLoughlin, Jr., *Modern Revivalism: Charles Grandison Finney to Billy Graham* (New York, 1959); Bernard A. Weisberger, *They Gathered at the River* (Boston, 1958).

30. Timothy Smith, *Revivalism and Social Reform* (New York, 1957); Winthrop S. Hudson, *The Great Tradition of the American Churches* (New York, 1953); Sidney Mead, *The Lively Experiment* (New York, 1963). Most of the essays in the last book had been published in the fifties, though some of their direction had been suggested earlier in Mead's *Nathaniel William Taylor* (Chicago, 1942).

31. Most of these essays are cited in Winthrop S. Hudson, "Shifting Trends in Church History," *Journal of Bible and Religion*, XXVIII (Apr. 1960), 235-38. For another list, see the section on "Religious Historiography" in Burr, *Critical Bibliography*, I, 22-27.

32. Something like this question is raised from the point of view of a church historian in the excellent article by William A. Clebsch, "A

New Historiography of American Religion," *Historical Magazine of the Protestant Episcopal Church,* XXXII (Sept. 1963), 225-57. Arthur S. Link argues eloquently that from the point of view of "Biblical faith," there is no such thing as Christian history as distinguished from other history. (Arthur Link, "The Historian's Vocation," *Theology Today,* XIX [Apr. 1963], 75-89.)

33. In the exuberance of the moment even Sidney Mead, sometimes criticized for his moderation by other "church" historians, said that it was no longer necessary to pay homage to "the rather presumptuous occupants of university chairs of secular history," or to orient church history according to "the unpredictable and transient interpretive vagaries" of these men. In 1963, however, he called for a much wider interpretation of the meaning of church history than these earlier statements suggested and condemned the tendency to widen the breach between religious and secular historians. (Sidney Mead, "Prof. [*sic*] Sweet's 'Religion and Culture in America' " [review article], *Church History,* XXII [Mar. 1953], 33-49, and "Church History Explained," *ibid.,* XXXII [Mar. 1963], 3-31.)

34. Martin Marty, *The Infidel: Freethought and American Religion* (Cleveland, 1961).

35. John Tracy Ellis, "The American Catholic and the Intellectual Life," reprinted in *The Catholic Church U. S. A.,* ed. Louis J. Putz (Chicago, 1956), 315-57; Henry J. Browne, "American Catholic History: A Progress Report, Research and Study," *Church History,* XXVI (Dec. 1957), 373.

36. Thomas F. O'Dea, *American Catholic Dilemma: An Inquiry into the Intellectual Life* (New York, 1958), 156, 158.

37. Walter J. Ong, "The Intellectual Frontier," in *Catholic Church,* ed. Putz, 394-415.

38. Oscar Handlin, "New Paths in American Jewish History," *Commentary,* VII (Apr. 1949), 388-93; Moses Rischin, *An Inventory of American Jewish History* (Cambridge, Mass., 1954).

· 5 ·

The Free Speech Movement
at Berkeley:
A Historian's View

The American campus revolution of the sixties began in
1964 with the Free Speech Movement at Berkeley. Some
people at Berkeley, including some conservative faculty
members, cannot help being perversely proud of this historical fact. For me as for many, the crisis was one of the
most difficult and one of the most educational experiences of my life. It profoundly shifted my ideas about
history, helping me to realize that the rather mechanical
style of detachment and neutrality that had long been
insisted on by most American historians could not, by
itself, get at the truth about movements of thought and
feeling.

Most of the original audience for the lecture consisted
of participants in the events described. For the benefit
of general readers I have added footnotes giving major
sources and explaining local references. I have also corrected one minor error of fact.

My principal written source was the workmanlike collection edited by Seymour Martin Lipset and Sheldon S. Wolin, *The Berkeley Student Revolt, Facts and Interpretations* (New York, 1965). Facts and quotations come from this collection unless otherwise identified. My most important sources however were my own memory and interviews with many participants.

This was my first effort to deal with this episode historically. I had earlier published two essays dealing with it as a contemporary observer, both in *The American Scholar,* respectively XXIV, 3 (Summer 1965), pp. 387-99; and XXVIII, 4 (Autumn 1969), pp. 588-605. The lecture was given in 1970 as the annual Moses Lecture at the University of California, Berkeley. It has not been previously published.

When Dean Elberg asked me to give the Moses Lecture, I asked what it was.[1] He answered, characteristically, by saying that it could be on anything and advising me to have some fun. I assumed, however, that even this generous invitation had some limits and that a lecture sponsored by the graduate division should be an academic performance in the lecturer's field. That is why I chose my present subject. My field is American intellectual history. I have worked in a number of periods. During the last ten years, my work in all periods has been affected by the fact that I have spent most of my time in the great revolutionary period of modern times—the late eighteenth century. Such has been the nature of life in Berkeley, however, that from time to time I like others have been forced to think, and even write, about the present period as well. I have chosen to give this lecture about my minor field, the immediate past, rather than my major one, the eighteenth century. In both periods I am specially concerned with the clash between intellectual enlightenment and various kinds of popular religion. I could thus say what I wanted in terms of either period. My purpose, in dealing with either, is to see things historically. This lecture is not intended as a manifesto or polemic, but as an appropriately academic performance.

When I say I want to look at this subject as a historian, this may not mean much. It would certainly not mean the same thing to all members of my department. What I mean is that I propose to look at it in the way I know best, with what equipment I have. This involves no very precise social-science technique. Some good work has been done on recent Berkeley by sociologists, and more should be, but my present method is an older and more impressionistic kind. Substantially, the method I am talking about means three things: first, one immerses oneself in the sources of a past time, trying to enter it empathetically and imaginatively as deeply as possible. Second, one comes back to the present and makes use of whatever vantage points the present affords. Third, one tries out whatever hypotheses the process has suggested, to see how well they serve to organize the data. This is not a method calling only for precise analysis of articulate ideas, but rather for a study of ideas and emotions embodied in actions and events.

Obviously such a process results in no final truth about any subject. At best it suggests some angles of vision. I would be very glad if my talk serves merely to start serious historical discussion of my subject. Because I think one essential for historical treatment of Berkeley in the sixties is met. The sixties are past, and in a sense more meaningful than that of the calendar.

Obviously, I make no pretensions in dealing with this subject to impartiality. Probably I cannot even tell you if I try what all my prejudices are. One is a belief, as a historian, that genuine relevance cannot be arrived at without some effort at detachment. Another is an ambivalent view of revolutions. There are always reasons for them, and sometimes they have to happen. They seldom turn out as anybody expects.

Americans, looking at the sacred revolutionary events of their past, the struggle for Independence and the war for the Union, are likely to go in first for undue reverence and celebration, then for sharp revision and muckraking. Both, I think, obscure reality.

The best and deepest history, even of an event sacred to many people, must try to get beyond either. A historian cannot aspire to any lasting success in dealing with such an event unless he makes a great effort to escape his prejudices, to entertain unwelcome hypotheses, to admit the facts that do and don't fit his theories, and to see the past in its own terms first, sharing as much as he can in its emotions as well as observing its events. The freshness that comes only from the sources is indispensable.

If, as I think, the double focus of an ambivalent point of view sees revolutionary events best, I have one advantage for my effort to look at my present subject. My emotions about the events I am dealing with have been both strong and thoroughly ambivalent from the beginning. This will doubtless become clear during my lecture: I will not bother you further with personal statements.

Historians have already decided that our university upheaval of 1964 is worth their attention. If you look at the most recent editions of quite a few textbooks of American history, you are likely to find the word "Berkeley" in the index. It will not refer to a pleasant California suburb, or to Nobel prizewinners. The same place-name, like Pearl Harbor or Sarajevo, even went into the language in the middle sixties as a common noun, as in "For God's sake, don't do that; we don't want a berkeley on our hands here."

The questions to ask about the event the word refers to seem to me the following: why did it happen, why did it happen *here,* and what did it mean? More precisely, why did a very prestigious university, in many ways the representative university of its day, become deeply convulsed, for a long time, over what seemed a minor dispute?

To answer this question, we must look briefly at the University in the late fifties and sixties, and even for a moment in an earlier period. In the 1930s the University of California was a pleasant provincial place. There was a lot of scattered excellence of a fairly traditional kind, and some mediocrity as well. Some of the most gifted faculty members chose to be in Berkeley because the way of life was more leisurely, the pace a bit slower than in the East. The East was a long way away; only a few professors chose to leave their families at Christmas and spend ten days going East and back on the train to take part in a national convention. There was a vigorous and diverse undergraduate life, in which football and fraternities contended with a growing fashion for intellectual and literary concerns and a healthy minority radical movement. The way teaching was carried on, which few questioned, was ordinarily in large lecture classes, in which one was periodically quizzed on the lectures and the textbook, sometimes written by the lecturer. The professors most widely known and remembered were platform artists with dramatic power and vigorous, even eccentric personalities. The ambitious undergraduate, however, could find his way in his last two years to a few small classes, some of them taught by gifted teachers under relatively little pressure to produce continuing evidence of research.

More or less this same university passed through the trials of the war and immediate postwar period, and in about 1950 was hit with special intensity by the Academic Revolution described by Jencks and Riesman in their book of that name.[2] One special California change was produced by air travel, which made our faculty fully part of their national profession, going East frequently to conventions, in some cases to Washington for government consultation. American culture in this period became more national and less provincial than ever before, and the government played a larger part in it. The power and pride and wealth of the United States, the prestige of its artists and scientists, made the fifties a great period of imperial culture, comparable with other such periods in other countries.

Along with nationalization came expansion both in numbers and money. The money came from three sources. One was the state, itself richer and doubled in population, carrying forward its generosity to education into a period of especially rapid expansion of the college-age population. The second was the federal government, which began making huge grants for science in about 1950. The third was the foundations, which reached entirely new dimensions. As the salaries of the faculty rose, so did the possibility of research support, leaves of absence, and world travel. As a part of the educational expansion, publishers sought out textbook authors with golden contracts, and were often willing even to take a chance on an important scholarly book.

A third change of this period is a more subtle matter, a change of ethos. According to S. M. Lipset and others, the United States veers back and forth between the two ideals of equality and achievement. This has certainly been true in terms of educational thought. And in this period, especially after Sputnik in 1957, the pressure was all for achievement. The mission given the university in a message of President Kennedy was to produce an "aristocracy of achievement arising out of democracy of opportunity." One of the later radical critics of Berkeley said about the same thing, that the function of the university had become the "massive production of specialized excellence." Professor Martin Trow said later it was to furnish a gate to the new meritocracy, a gate which would let some in and keep others out.[3] This was an important assignment, on the whole well done, and very characteristic of the whole society. At the time, what admirers of the new United States emphasized was the increased prosperity and mobility, the talents the nation was able to train and mobilize. People had not yet noticed that a great many were left out.

The academic revolution affected all parts of this university, administration, faculty, and students. As an administrative problem, the university had become analogous to the United States. A vast administrative machine had been created, with brilliant top direction and a cumbersome middle bureaucracy. The University of California with its nine campuses, and the state system with its three levels, was admired throughout the world, and imitated where resources permitted. The cover of Clark Kerr's book[4] quotes Noel Annan in the London *Sunday Times:* "Clark

Kerr runs a university with seven campuses (including Berkeley, the greatest campus in the world). . . . Reading this book is like listening to someone living half a century later than we are."

The faculty amounted, perhaps, to the greatest concentration of academic talent ever assembled in one place. It was expanded very rapidly, and on the whole the recruiting was done carefully and on the basis of merit. Faculty members worked with great intensity to make the university, or at least their department, the greatest anywhere.

The whole picture, seen from where we are, cannot help arousing a certain amount of nostalgia. Yet this great vigorous institution had, like the United States itself, the defects of its virtues.

An essential quality of the scholarly production of the time was its insistence on detachment, which is indeed in some sense a necessity of scholarship. It was the day of highly technical literary criticism, of value-free sociology, and of a revolt, in American history, against the easy progressive-conservative or good and bad dichotomies of the age of V. L. Parrington. Scholars placed a high value on subtlety and complexity. Just as many sociologists and social psychologists regarded the resolution of conflicts as the norm, historians tended to see much fundamental agreement among American antagonists of the past. Anti-intellectualism, simplification, mass emotion, seemed the main enemy. John Higham, a leading intellectual historian and also a stern moralist, writing about his colleagues in 1962, after granting what he called the "consensus historians" many virtues, had this to say about recent developments:

> In functioning as a conservative frame of reference, the consensus approach gave us a bland history, in which conflict was muted, in which the classic issues of social justice were underplayed, in which the elements of spontaneity, effervescence, and violence in American life got little sympathy or attention. As the progressive impulse subsided, scholarship was threatened with moral complacency, parading often in the guise of neutrality.[5]

It is wrong to think of the faculty of this period as uninterested in teaching. Most were very hardworking and conscientious, and quite a few much concerned to raise the level of instruction, especially its top level. There was a new emphasis on honors programs, an effort to establish small senior courses. There is no question, I think, that graduate and upper division teaching immeasurably improved. Yet the basic problems inherited from the earlier period were not solved, enormous lecture courses remained, and nobody respected any more the old type of platform artists.

In choosing new faculty, there was a great effort to comb the nation and sometimes the world for excellence. This usually meant excellence in writing. When people said we need a man in this or that field, they nearly always meant no research is being done on this campus in that field. When they said such a man would greatly strengthen the department, they meant, of course, that he would raise its prestige among similar departments in the nation.

In department after department, there was a vigorous attack on slipshod standards. Much of this was healthy, but sometimes it seemed to involve toughness for its own sake: sometimes the assumption seemed almost to be that the more students thrown out, the more graduate students failed, the more assistant professors dismissed, the better the university would automatically become.

Despite their new comparative affluence and their pride in their work, my memory is not of a happy faculty. To some extent, all American intellectuals suffered from the tensions of the recent past, from the intellectual flirtation with Marxism in the thirties, followed by revulsion against the crimes of the Stalin regime. Some felt uneasy about their own behavior during the McCarthy period, though many had spoken out against repression. In Berkeley, the faculty suffered from the deep trauma of the Loyalty Oath, in which the faculty victory took the odd form of spreading a compulsory oath to all employees of the state.

Even more than from such problems, faculty members suffered from fatigue, brought about by the demand for intense concentration on research, in the midst both of teaching and of complex administrative duties. Such tension showed itself in intense reaction to all issues raised, to the problems of parking, for instance, or the much more serious one of the introducton of the quarter system, in which the faculty proved to have no power. All the tensions of the system were greatly increased by a speeded-up schedule, justified in terms of full utilization of plant. Many faculty members in this period went for medical treatment to Kaiser Oakland. There some of the more knowing doctors, after hearing the complaint of routine symptoms of nervous tension, would glance at the patient's card and say "Ah yes, you are at the university." In many ways, American professors and American businessmen had come to resemble each other more than either would like to have admitted.

The students were also affected by the climate of competition and meritocracy. They were on the whole hard-working, until about 1960 generally uncomplaining, and always polite to the faculty. According to one of our best-informed observers of student population, a major change began to take place in 1960. Like the faculty, the students increasingly were drawn from all over the country and the world. After long complaints of apathy, students began, with the whole society, to discover new problems in America, or rather old unsolved problems. From about 1958 the existence of urban poverty had become increasingly realized. The civil rights movement engaged student emotions, drawing some first to the dangers and heroisms of Mississippi, others later to the sit-ins of Berkeley and San Francisco. At least as important, students began to experiment a little with new lifestyles, to take an interest in the Beat movement in San Francisco. The churches on Dana Street, which had been surprisingly full of students in the fifties, lost them, as the young sought elsewhere for value and meaning.

As for the administration, the university as seen from its vantage-point is brilliantly described for all time in Clark Kerr's Godkin Lectures, delivered in 1963 and published just before the deluge in 1964: surely the most penetrating, the least discreet, and the least pompous book of the kind ever written by an American university or college president.[6] As one reads it now, though it is in part a celebration of the university's new complexity and new greatness, it is also full of sad wisdom and rueful wit. It sees the distortions and lack of control inherent in the huge grants, however well intentioned. Its famous description of the faculty as a group of entrepreneurs loosely tied together by a grievance about parking is more than funny. It makes the quite extraordinary statement, much stronger than most made by outside critics, that none of the changes and improvements of the period had greatly benefited the undergraduate students. It says that the university despite its excellences has no single soul to call its own. It predicts a student revolt. Above all, one has in reading it, as in reading memoirs of Kennedy's national administration, a sense of delicate and precarious balance, making quick movement very difficult.

Finally, though Kerr praises, and rightly, the public of the state for its generosity and increasing tolerance, it seems now as though the great university of the fifties and early sixties, with its national and international reputation, its vigor, its confidence, had grown out of touch with its California constituency. Doubtless the people of the state were proud of the Nobel prizes and the plaudits. But most alumni feeling was centered on the old university, which no longer existed. The new one was impossible not to admire, but it was not easy to love, and that turned out to matter.

It is not necessary to recall to you in any detail what happened in 1964; the facts are set forth with admirable lucidity in the chronology prepared by our alumni publication, and included along with many documents in the Lipset and Wolin volume on the *Berkeley Student Revolt*. I want now to make only one chronological point, that the Free Speech Movement came in two phases. The first was a complex dispute over the right to collect money and signatures for candidates on a strip of land which seemed to be outside, but proved to be inside, university property. This was quickly settled in favor of the students, but gave rise to a general dispute over political activity on campus, escalated through a number of minor sit-ins and sporadic attempts at discipline. The issue, by December 1964, had come down to one major unresolved disagreement, over the right to advocate on campus off-campus action which might prove to be illegal.

But this statement gives no idea of how it really was, and the history of the crisis will have to be written by another kind of historian than a constitutional expert. What we remember is the hot day in Sproul Plaza, with the students sitting around the police car doggedly and sadly, hour after hour, the certainty of impending disaster, the jubilant announcement of agreement with the president.[7] We remember Mario Savio urging students to put their bodies on the line, or Joan Baez urging them to enter Sproul Hall with love in their hearts. We remember the banners of liberation flying from Sproul Hall and Sather Gate and then the announcement in the middle of the night that students were being removed by police. Above all, some of us remember the announcement in the Greek Theater, after much hard work, of an agreement between department chairmen, president and regents which would usher in a new period of freedom and order, and the immediate collapse of the campus into indescribable chaos. We remember the adoption then of a resolution by the faculty taking substantially the student position, with the euphoria of the majority and the bitterness of the minority.

I wish to make the point here that the tone of the movement changed in the middle of the controversy, and that the issues moved into an area where nobody could grasp or understand them completely. And I want to make it especially clear that this does not imply a dismissal or derogation. To say that this movement, or any other, carried some emotional freight which its leaders did not know was aboard, is to make it more, not less

significant. To anybody who looks at the sources or talks to participants or consults his memories, it is clear that the rhetoric of
the movement began to change about the time of the car incident, and had changed completely by December 8.[8] Instead of
complaints of particular grievances, or complex arguments in
terms of constitutional rights, students began talking about a
glimpse of the university as a loving community, or finding a
moral meaning for their lives. Sudden conversion was a common
phenomenon. The ASUC Senate was against the movement six
to one until after December 8, when a new election made it all
pro-movement, and the outgoing president said "We've missed
the boat." Many faculty members took quite different positions
in the crisis than they did shortly before or after.

To account for all this needs many kinds of history. I will summarize briefly the main theories about the nature of what happened. I will move from the more to the less concrete, and thus
in my opinion from the less to the more important.

The easiest immediate explanation was to blame the crisis on
administrative mistakes. There were plenty of them, as President
Kerr admitted only a little later:

> We fumbled, we floundered, and the worst thing is I still
> don't know how we should have handled it.[9]

Like the United States government, with its eighteenth-century
checks and balances, the university government, with its divided
authority and complex structure, works well in times of consensus and calm, and is strained by crisis. It seems clear now that
even if the administration had reacted to this new and devastating challenge with godlike compassion, understanding, and firmness—which it did not—there would have been a similar outbreak
in some other university, and the movement would have spread
to Berkeley instead of from Berkeley.

The next easiest is to make it a radical plot. Of course there
were radicals who wanted something like this to happen. But I
am convinced that the course taken by the FSM went out of their

control and was successful beyond their dreams. In the next five years some radicals got a lot of skill in manipulating the three-stage scenario of outrage, police intervention, and mass protest. But there were far more attempts than successes. Unless the cause at stake caught the volatile emotions of the students and non-students, the scenario failed. Which cause would catch these emotions and which would not was unpredictable: the only crisis comparable in scale to that of '64 (and one far more violent and tragic) arose out of the largely unforeseen and, to many people, baffling issue of the People's Park.[10]

Third, the episode can be seen as a struggle over free speech, which of course in part it was. With the original issues quickly settled, the dispute came down to two conceptions of the relation between the university and political action. The administration position was that the university must be free but partly aloof. Students had just acquired the right to have speakers of any kind (including Communists who had previously been prohibited) speak to them in university buildings under certain restrictions including the right to answer. For this advance President Kerr received, in April, the Alexander Meiklejohn Academic Freedom award from the American Union of University Professors. In the administration's view, in return for this freedom, greater than that actually then available outside the university, the university had to be non-political, and direct political action as opposed to political speech had to be prohibited. The FSM position, repeated again and again, was that there should be no difference between political speech and advocacy on or off the campus, that students should be treated as citizens, subject only to constitutional restraints, amenable only to the courts.

This is not, I think, a simple issue, and after some drastic samples of what it can mean to be subject to outside civic authorities some may have some respect for a degree of aloofness. The administration position was, however, an impossible one to maintain under the circumstances. The Dean of Students at one stage of the argument, trying to be very conciliatory, defined the university position as follows: "a speaker may say, for instance, that there is going to be a picket line at such-and-such a place, and it is a worthy cause and he hopes people will go. But, he cannot say, 'I'll meet you there and we'll picket.' " It was quite impossible that emotionally aroused students would carefully observe such a fine distinction.

The dispute could be seen as an extension of the civil rights struggle. Some leaders of the FSM had been involved both in the Long Hot Summer in Mississippi and in the more radical northern movement, in Auto Row and the Palace Hotel. It was widely believed in the movement that the administration and regents, in response to pressure, were trying to prevent further local sit-ins. Whether or not this had any truth in it I have no idea, but the belief made it impossible to settle the issue, as some of us tried to, on the basis of amnesty and postponement of the "mounting" issue.[11] To some of the students this last was the heart of the matter. At the time, the whole legal status of the sit-in seemed to some to be ambiguous: southern sit-ins to secure legal rights were compared with sit-ins intended to force one's opponents to capitulate. This issue was settled by the courts in the aftermath of the FSM: mass sit-ins were not a legal method of protest and were seldom tried again.

Some have said that the cause of the upheaval was educational discontent. The FSM itself spoke of the educational factory, IBM machines, etc., and the movement soon discovered the issue of student power. Curiously, much the same rhetoric became customary for conservatives. The conservative press, Governor Reagan, President Hayakawa,[12] even sometimes President Nixon, in denouncing student radicals, were likely to throw in a paragraph about undue concentration on research and graduate work, large lecture courses, leaving too much teaching to untrained graduate assistants, and the need for a student voice in education. Yet still more curiously, surveys made at the time of the FSM found little student discontent with Berkeley teaching. During all the efforts at educational reform in succeeding years, it proved difficult to get any substantial expression of student opinion and still harder to get positive suggestions for change. And many students, many of them excellent, continued to transfer to Berkeley from other institutions, often from institutions widely praised for innovation and experiment.

So much has been said about teaching here that is uninformed, critics have so much neglected such matters as size and the complex assignment of the university, so many efforts have been overlooked, so many ultra-simple answers suggested, that we are often tempted to go in for angry and self-righteous defense. Faculty really know that there is some substance in the principal criticisms. As for student opinion, the fact that there have been few constructive suggestions and much ambivalence does not mean that students are content. And students, faculty, and particularly the outside public are all affected by another of the mysterious pendulum swings of American educational opinion. Throughout the whole system there is a reaction against the values so vigorously espoused in the fifties. Instead of tracking, honors programs, strict standards for admission, the demand is for more flexible criteria, the use of education for reform, and a downgrading of research. I question whether this swing had quite begun at the time of the FSM; most of its activists seemed to handle the existing educational program successfully.

Coming now to the more controversial and I think more interesting suggestions, the one that was most unpopular at the time was that the upheaval was a generational battle. Yet surely a generational conflict was going on, and of a far more crucial nature than anybody realized. Part of this was a wide and general demand for a lowering of the age of majority. Without any clear signal being given, middle-class youth demanded that complete equality be granted at about 18. And they have largely won this battle. Everywhere regulation of students' private lives, by parents or colleges, has broken down, and in particular the sexual mores of the previous period have found few consistent defenders. The voting age will probably soon be lowered.

Much more important than the demand for majority at 18, or than any negotiable demands, was the emergence of the new youth culture, and the relation between this and the FSM is probably the most important and baffling question for future cultural historians of the episode. Personal liberation, emotional self-expression, the search for identity: none of these I think was the purpose of the movement when it began, and some of the radical leaders of that day had little use for this whole vocabulary. The movement ran away from them; in its course, participants have told me, new needs and hopes were discovered.

I shall not attempt a description of this major, irreversible change at this point. At the time, the youth culture was described best by Richard Herr's essay on the students in the opening section of the Muscatine Report,[13] a sympathetic yet critical view of the demand for "instant poetry, instant psychoanalysis and instant mysticism." This account saw behind the gaiety and the need of collective self-expression a potential for withdrawal, even despair. Some others at the time saw the movement as part of a widespread collapse of the dominant morality, related to changes running all the way from demography to religion. I will quote just one statement, that of Laurence Veysey, a young historian now at Santa Cruz, a strong statement but not I think a silly one:

> Perhaps the most fundamental rift between Americans is no longer that running between white and black, but instead the one running between those who believe in planning, large-scale organization, a conventional work and family life, and the rights of private property, and those who believe in decentralization, separateness, spontaneous expressiveness, and the prime value of survival as such.[14]

Finally, and this perhaps is another way of saying the same thing, the FSM coincided with a dawning realization in America that the progressive triumphs of the New Deal, Fair Deal, New Frontier, and Great Society, were not a success. This change has recently been described as the dawning of Consciousness Three.

If one wants to put it negatively, instead of the Greening of America, one might call it the Souring of Progressivism. As is often the case in this country, the values of the immediate past were repudiated with undue sharpness. The Civil Rights Movement seemed to have accomplished nothing for the ghettoes, urban renewal seemed the uprooting of the poor, the rise in the standard of living in the suburbs had merely deepened the miseries of the center cities, the defense of freedom had become in the eyes of many people aggression bordering on genocide,[15] and, most devastatingly of all, the triumphant rise of the GNP had meant the destruction of the environment. Causes to which liberals had given their energies and devotion for a generation seemed a failure. And this is my principal explanation for why it happened at Berkeley. This university, a triumph of organization, of competitive effort, of upward mobility, of productivity, this great concentration of specialized excellence, represented the epitome of American culture of the 1950s at its best; and it was exactly that culture that had lost the confidence of the young.

Historians trying to make a general point often make use of analogies. Since human beings in different periods have many feelings in common, analogies are often illuminating, though always inexact. Certainly every historian I know, in 1964, called on his specialty for precedents, and every specialty from ancient Greece to modern Germany furnished telling similarities. Some talked in terms of revolutions. The FSM, or the student movement in general, was not a revolution, since no revolutionary government emerged, no new institutions were created. Yet at least one point in the dominant interpretations of the French and American Revolutions seemed relevant: both had started as a vindication of existing constitutional rights, and both had proceeded by a series of stages farther than any—or almost any—of the early rebels had expected. The words of Robert Palmer, written in 1958, about the pre-revolutionary situation of Europe in the 1760s, are not without meaning either for the United States or the campus two hundred years later:

By a revolutionary situation is here meant one in which confidence in the justice or reasonableness of existing authority is undermined, where old loyalties fade, obligations are felt as impositions, law seems arbitrary, and respect for superiors is felt as a form of humiliation; where existing sources of prestige seem undeserved, hitherto accepted forms of wealth and income seem ill-gained, and government is sensed as distant, apart from the governed and not really "representing" them. In such a situation the sense of community is lost and the bond between social classes turns to jealousy and frustration. People of a kind formerly integrated begin to feel as outsiders, or those who have never been integrated begin to feel left out.[16]

Another analogy, a little closer I think, compares the current movement to earlier intellectual rebellions. The two chief specimens of the kind in the United States began in the 1840s and '50s, a period of triumphant Jacksonian democracy; and in about 1912, a period of triumphant progressivism. In both, young intellectuals, especially young literary intellectuals, demanded kinds of freedom and kinds of change that had nothing to do with the progress of which their elders were so proud: instead of democracy or economic growth they demanded a richer and freer life of the spirit. Perry Miller, our greatest intellectual historian, describes the Transcendentalist movement, in 1950, in words which surely have a new resonance:

The protest of these few troubled spirits against what their society had confidently assumed was the crowning triumph of progress and enlightenment is therefore a portent for America, all the more because their protest was the result of no organized indoctrination, but was entirely spontaneous and instinctive.

The movement, he says, was "nothing less than the first of a succession of revolts by the youth of America against American Philistinism."[17]

Yet these revolts of young intellectuals, fascinating and intellectually powerful, important in their long-run effects, remained small elite movements while the revolt ushered in by the FSM

has spread through a large part of a generation. The most helpful comparison drawn from the American past, I think, is that to
religious revival. In invoking this metaphor, I am by no means
trying to make the FSM into a group of foolish fundamentalists,
nor am I trying to make it 100 percent pure, holy, or beautiful.
Revivals in their greatest periods in America, were movements of
irresistible power. One of them helped to bring about the antislavery movement, much as the student movement of our day
tried to achieve a connection with the quite different Black revolution. Revivals start with a conviction that one's life and its values are unsatisfying; they progress through despair to brief euphoria. Their method, at their most successful, is not argument
but drama or witness, we might say demonstration. To their
opponents they are violent, obscurantist, and above all anti-
intellectual, and to their adherents these opponents are cold-
hearted and rigid formalists. Once revivals get going, they spread
very fast, yet one way or another they also end soon; the emotional pitch they demand cannot be sustained. Sometimes they
are taken over by different people from those who start them and
successfully embodied in institutions—as the movement I am talking about has not been. Sometimes they simply peter out. In the
process, as their opponents always point out, they often run into
all sorts of extravagances, blasphemies, nudities, promiscuities,
even crimes. People who believe they have a divine mandate,
known only to their hearts, are often dangerous people. Yet even
when revivals are over their effects remain, sometimes as a re-
vitalizing of old values, or an opening to new ones. They are remembered as stirring times; even some who never surrender to
them are conscious, to some degree, of a change of heart, or shall
we say a change of consciousness. Institutions almost always find
them impossible to contain or comprehend.

Analogies aside, the FSM won its immediate goals. The larger
movement it inaugurated, however, seems to me to have been,
like Transcendentalism or many revivals, institutionally a failure
and emotionally a success. Among its consequences have been the

following: For a while at least it made the students another center of power, parallel to the legislature, the regents, or the faculty, with a veto on university action they had not had when Clark Kerr wrote his book. Yet the students did not prove able to use this new power effectively. Demonstrations proved undependable, particularly with the sharp rise of public hostility. Strikes were unsuccessful, with the partial exception of the Third World strike for an ethnic studies department, which called in a new constituency.[18] As for outside political participation, despite some local successes it seemed very difficult to mobilize enough sustained and realistic effort, and the students did not succeed in finding many outside allies. The only educational changes that came about, aside from the ethnic studies department, were made by faculty, who asked for far more student participation than they got. These changes included some experiments in student-sponsored courses and summer colleges, some of which turned out well and some badly, the change from grading to pass-fail for some courses, and finally the end of the breadth requirements of the college of letters and science. Most of these changes could, I think, be related as much to the swing of educational thought in the country as to student pressure. So far, I think, there have been no solutions to the major structural problems of the university, inherited from several decades ago, and little progress in redefining liberal education or giving it unity and coherence. The most conspicuous venture in that direction, and the one most closely related to the FSM, was the Tussman program,[19] which failed for lack of either faculty or student support.

Thus it is hard to close on any but a rather somber note, and I have no wish either to be bland, or to romanticize a partly romantic movement. We all know how grave the situation is in which we now find ourselves, with so many and diverse critics and opponents that survival is in question.

Yet I am not willing to end on that note either. What has been most important for Berkeley is the cultural change that spread and grew far beyond Berkeley, making Berkeley for better or worse one of its major centers. This new way of thinking and feeling spread far more rapidly than that of 1840, or 1912. A graduate student from this period has said recently to me that Berkeley in the second half of the sixties was schizoid. On the one hand it remained the university of the fifties, rather conservative academically, with demanding intellectual standards. On the other hand, Berkeley became a center of the new culture, permissive, emotional, and expressive. To me, it is out of this division, this contrast, that all that is best in Berkeley of the late sixties has grown.

Between our crises, despite the anti-intellectualism of the surrounding hippies and the rising hostility of the public surrounding *them,* without much institutional change, teaching here has been a lot more interesting recently than it was in 1960. At its best, the combination of the old and new Berkeley styles has been more fruitful than either alone. Professors have been forced to re-examine the complacent assumptions of the fifties, to examine new possibilities, to answer new criticisms. Sometimes, I think, the inward search of the young has been successfully related to complex reality on several levels. Sometimes the moral fervor of young radicals has been deepened, not blunted, by a challenge to think things through. At best, the creative confrontation between the academy and the movement has been useful to both: *it only can be if each maintains its own character and function.* If sometimes this has given our classes the character of combat, it has usually been friendly combat, and I have the feeling that its rules are becoming better understood: neither side demands surrender, only honesty.

I make no prediction; that is not the business of historians. But if our mission in the seventies is to continue this creative clash of styles, and if we are permitted to pursue that mission, Berkeley in the seventies, like Berkeley in the sixties, will be one of the most interesting places in the world.

Notes

1. Sanford S. Elberg was dean of the graduate school. The Moses lectureship was endowed by the President and Regents in 1904 to honor the memory of Professor Bernard Moses.
2. Christopher Jencks and David Riesman, *The Academic Revolution* (New York, 1968).
3. I regret that I am unable to find the source from which I quoted Professor Trow's remark, which seems to me accurate and acute.
4. Clark Kerr, *The Uses of the University* (Cambridge, 1964), paperback edition, New York, 1966.
5. John Higham, "Beyond Consensus: The Historian as Moral Critic," in *Writing American History* (Bloomington, Ind., 1970), 146.
6. Kerr, *Uses.*
7. In this incident (October 1, 1964) the police car contained a former student, arrested for trespassing while soliciting political funds contrary to existing rules. Students immobilized the car, and police gathered, prepared to use force to liberate it. At the last minute a violent confrontation was prevented by a hastily negotiated faculty-student-administration agreement.
8. This date, remembered as historic by supporters of the student position, was that on which the Academic Senate came round to essentially the student position on the regulation of speech and political action.
9. Quoted in Lipset and Wolin, *Revolt*, 430.
10. In 1969 a dispute arose over the use and control of a vacant strip of land owned by the university and adapted as a park by students and others. In the ensuing dispute one person was killed, another blinded, and Berkeley was briefly occupied by fully armed national guard units.
11. The "mounting issue": the problem of speech *on* campus that gave rise to violent or illegal action *off* campus.
12. S. I. Hayakawa, then president of San Francisco State University, later Senator.
13. The Muscatine Report was a set of proposals for educational reform at Berkeley drawn up by a committee headed by Professor Charles Muscatine. It was published in 1966 by the Academic Senate under the title of *Education at Berkeley*. The Herr quotation is on p. 30.
14. Laurence Veysey, *Law and Resistance* (New York, 1970), 5.

15. This phrase about the Vietnam War is slightly anachronistic. The escalation of the war, which later provided the student movement with its most telling issue, had barely begun and played little if any part in the events of 1964.
16. R. R. Palmer, *The Age of the Democratic Revolution*, vol. I, *The Challenge* (Princeton, 1959), 21.
17. Perry Miller, *The Transcendentalists* (Cambridge, 1950), 8.
18. This, like the People's Park incident, took place in 1969.
19. A small experimental residential college program, organized by Professor Joseph Tussman in 1965.

. 6 .

The Problem of the American Enlightenment

In the sixties I started to put together a book on the Enlightenment in America. At the same time, I was becoming increasingly interested in religious history. To some, this combination seemed incongruous, but I became convinced that the Enlightenment was in essence and origin a change in religious opinion. Because of the complexity of the subject, and also because this was a busy time of campus crisis and administrative commitment, the book was not published until 1976.

This article was an early attempt to organize the topic, largely on a chronological basis. As the article suggests, however, I found it necessary to emphasize the different *kinds* of Enlightenment in order to understand the spread from Europe to America of Enlightened ideas. In the book I shifted much farther from a chronological to a thematic organization.

The last paragraph shows clearly the influence of the campus crisis discussed in the previous essay.

This essay was presented as a paper at the 1967 meeting of the Organization of American Historians and published, after considerable revision, in *New Literary History,* in a special volume on periodization, I, 2 (Winter 1970), 201-14. It is reprinted by permission of the editors.

The problem of the American Enlightenment is full of paradoxes. The first of these is that while most American historians are its partisans, only a few choose to write about it. The opponents of the Enlightenment, Calvinists and revivalists and romantics, have been favorite subjects of our intellectual historians and literary scholars. Many historians, it is true, have written good books about some aspects of the Enlightenment: its social background, its science, and especially its great political figures. Yet few have made much effort to analyze its philosophical or religious allegiances, fewer still have tried hard to compare or relate it to the various European Enlightenments, and nobody at all has attempted anything like a quantitative enquiry into the spread of Enlightened ideas among Americans. One result is that general historians allow themselves to make in passing all sorts of statements that seem—perhaps more than they really are—contradictory. The Enlightenment met little resistance in colonial America, yet was mostly an upper-class affair. The founding fathers were mostly deists, yet the Calvinist clergy were the most consistent partisans of the Revolution. Some sharper contradictions have had less attention than they deserve. When our only real *philosophe* was elected President in 1800, evangelical revivals were sweeping sections of the country that had voted for him.

One cannot blame American historians too much for their failure to solve these paradoxes if one looks at historians of the European Enlightenment, particularly the best of them. To Paul Hazard, the great clash of ideas had taken place, and virtually all the views of the Enlightenment had been stated, before 1715.[1] To Peter Gay, the Enlightenment was materialism or scepticism; its best exemplars Diderot or Hume; all of early modern thought was its prehistory.[2] To Ernst Cassirer everything pointed toward

Kant; to Leslie Stephen everything seemed to be a reaction to the English Deists.[3] Dating and defining the Enlightenment so differently, historians have inevitably made different judgments about its consequences. To a host of nineteenth- and early twentieth-century historians, it was liberation from superstitious darkness; it paved the way for positivism, or Marxism, or democracy. To a few, perhaps differing mostly in their perspective on their own time, it was the beginning of breakdown, issuing inevitably in intellectual chaos, totalitarianism, or the revolt of the masses.

Obviously a historian of the American Enlightenment cannot settle all these questions briskly before he goes to work. Equally obviously, he needs some tentative and working answers to the most obvious questions. The answers I am about to suggest are those I find most tenable at the present stage of my work on the subject. The necessary questions seem to me these:

1. What is the Enlightenment?
2. Is there an American Enlightenment?
3. If there is, how is it related to its European counterparts?
4. How widespread were Enlightenment ideas in America?
5. What is the period of the American Enlightenment; does it have clearly marked beginnings, stages, or end?
6. Finally—since all Americans, enlightened or not, make didactic use of their history—what of it? How much does the American Enlightenment matter to us?

If we were to start with as rigorous an answer to the first question as Peter Gay's, we might have to give a nearly negative answer to the second one: real scepticism or materialism were extremely rare in early America. We can, I think, make our definition more general without having our subject evaporate. Such shape as it has it takes partly from contemporary consciousness. Many Americans living in the late eighteenth century believed that they lived in a period more enlightened than those which preceded it. And I have run into a number who sensed the end of this period, people who felt themselves uncomfortable in the age of Bible societies, sabbatarian movements, and emotional pre-Jacksonian politics. These are not all stuffy old gentlemen of what Parrington called "the tie-wig school"; they include for instance both Jefferson and Adams.[4]

What Americans of the late eighteenth century meant when they talked about "this Enlightened Age" is not, I think, hard to formulate if one does not insist on too precise boundaries. They meant that it had recently become possible, through the proper use of the human faculties, to understand the Universe better than it had ever been understood before, and to make practical use of this understanding. Not all of them believed that light would spread equally over all fields. Some of them thought that moral philosophy was likely to remain a little cloudier than natural philosophy. Most believed that some topics were best dealt with by Revelation—for instance life after death. Not all were sure that progress would continue forever. Many seem to us to have dealt unsatisfactorily with the question—central to them—of the precise faculties and methods through which progress was being made. They were almost all Lockeians and, like their master, many found the road from external reality to sensation and reflection a more difficult one than it had seemed when they set out. In America, even more than in Europe, we have to define the Enlightenment partly as a mood or a style, and to state its common content in the simplest terms. If we say merely that Enlightened people believed that the universe was comprehensible by reason and becoming more so, this does not define the Enlightenment so broadly that it has no boundaries left: to some Americans the universe remained finally mysterious, understandable only through divine illumination; to many revelation was still the alphabet rather than a mere teaching aid.

If we define the American Enlightenment this broadly, we have almost answered the second question, whether it exists. Yet this question has been raised so persuasively by Daniel Boorstin that we must pause a moment to answer him. Mr. Boorstin says that the idea of an American Enlightenment is an illusion foisted upon us by "intellectual historians."[5] Though Mr. Boorstin is sometimes himself a very illuminating intellectual historian I think he is wrong here. To say that there is no American Enlightenment must mean one of the following things: that there is no periodization possible in the history of thought, that the European Enlightenment is itself a delusion, that American intellectual history is quite separate from European intellectual history, or that Americans are not interested in ideas. I cannot accept any of these statements. To say that people differ about the definitions and dates of the Renaissance or the Romantic period—or the Enlightenment—does not mean that these terms cannot be used at all, rather that historians using them must define their terms more carefully.

Similarly, to say, as Mr. Boorstin says well, that Americans think in some ways differently from Europeans and are deeply affected by living in a somewhat different society does not really take care of the question of European origins. One can hardly deny that American culture is far more closely related to Western European culture than to any other. The American Enlightenment, like American Romanticism or for that matter American Christianity can be at once American, Anglo-American, and European; one does not have to make a general choice but to discriminate among different elements.

There is one more possible meaning to Mr. Boorstin's denial, one which he argues boldly where others have hinted. This is that Americans are simply not interested in any general ideas; that European intellectual history exists but American does not. This lack of ideology is sometimes linked with democracy; rather oddly, ardent democrats have implied that general ideas are things the plain people don't have. It seems to me one can argue equally well that the least educated and the least articulate sometimes have the strongest and most dogmatic beliefs; only a few people, as Voltaire pointed out, have the time and leisure to be sceptics. Often enough, historians have said on one page that

Americans are hard-headed and practical and on the next that they are obsessed by moral simplifications or political generalizations. Both these statements are right, and pushed far enough they may come together. Americans are as interested as other people in general ideas, but in a somewhat different way. On some things their consensus is so strong that it is tacit and unconscious. In pointing this out—and this I think is his central meaning—Mr. Boorstin has done his greatest service of all. His work is not the only instance of important historical progress through creative exaggeration.

We come now to a more difficult question, and one I can deal with only very sketchily here, the question of the differences and similarities between the American Enlightenment and the Enlightenment in Europe. The causes of these differences seem to me a little simpler than their definitions, and I would group them in two large categories: the environment of the animal and its natural enemies.

The first is familiar, and I will not try to add to all that has been said, by Mr. Boorstin and others, about the relative egalitarianism, expansiveness, diversity, and practicality of American society. Perhaps a word of caution is in order about diversity: in terms of religious allegiance or national origin American society was indeed the most diverse in the Western world, but in terms of social extremes—with one tragic exception—the range was smaller. There were few noble or very wealthy patrons to sustain or insult our intellectuals, few bright but alienated Grub Street hacks, few full-time intellectuals of any kind. The American Enlightenment had to depend on such busy people as planters, doctors, lawyers, politicians, and not least preachers. And these were constantly complaining, in the eighteenth century as in the nineteenth, about the *relative* lack of libraries, publications, learned associations, and leisure. Something they mentioned less often was the lack of mathematics and languages. Few were equipped to make fresh contributions to physical science. It is clear that many read some French, but it is very hard to tell how fluently and easily. It is arresting to learn that Rittenhouse had to learn Latin specially to read the *Principia;* would not this limitation have been rare in Europe among otherwise informed potential readers of Newton?

Yet all these disadvantages are most striking when one compares America and Europe. Perhaps it would be more reassuring, as well as more modest, if we compared America in cultural achievement to other colonies of Europe. Then the question might be, why did this particular colonial civilization make so early, so individual, and in some ways so impressive a contribution to the common culture?

To return to the sources of difference, one of the principal advantages and also one of the principal disadvantages of the American Enlightenment was its lack of an *infame* to *écraser.* Unlike France, America had no old, unified, official establishment, extending from religion into both science and politics, armed with full powers of censorship which it used inconsistently, exactly dangerous enough even in its decadence to make it the perfect enemy. Unlike England, America had no tolerant, assimilative establishment, able to point with satisfaction to a recent revolution and say that progress was obviously in good hands and going fast enough.

Of the possible enemies for an American Enlightenment, the most obvious, Catholicism and absolute monarchy, was indeed an important national enemy up to 1763 and an equally important cultural enemy long after. The assumption that progress took place *against* the world of Philip II or Louis XIV was almost universal among Anglo-Saxon Protestants for a long time. While the French and Catholic threat was real and present, before 1763, it was clear that Protestantism, England, and Enlightenment were on the same side; only later—and perhaps only temporarily—were these allies separated in the dominant American view of the past.

It is true that during the Revolution and for some people

afterwards England became the corrupt and aristocratic enemy of progress. In New England, particularly, as Mr. Bridenbaugh and others have shown us, Anglicanism as well as monarchy, mitre and sceptre, were seen as forces of tyranny and reaction, and some of the old anti-papist weapons were turned against them.[6] Yet it is difficult for modern scholars to place the Enlightenment, as we have defined it, clearly on one side or the other of the struggle between New England orthodoxy and Anglicanism. Indeed, some of the black-coated regiment which loyalists saw as their fiercest enemy feared the Arminian looseness of the Church of England as well as its hierarchical organization. The Crown, which had two generations earlier forced the toleration of Quakers and Anabaptists as well as Anglicans, was now extending its hypocritical generosity to the Roman Catholics of Quebec. And the latitudinarian Church of England could hardly be cast in the role of religious rigidity: too many sinners, disciplined by the First Church, had found ready acceptance across the green at St. George's.

A very common suggestion is that Calvinism was the chief American enemy of the Enlightenment. Both Jefferson and Madison sometimes cast the New England clergy in this role, and in Jefferson's private correspondence Athanasius and Calvin were major villains of mystification. A great many courses and books in American intellectual history solve their organizational problems by contrasting "Puritanism" with Enlightenment. In some ways the task is more difficult in relation to America, since American Protestantism was so extremely diverse. It used to be assumed that Protestantism was one stage on the way from the Middle Ages forward to rationalism. This is hardly any longer the view of historians, though it has wide popular currency. In America, the doctrine of salvation by unmerited grace that was defined by many as the consistent center of Protestantism was, as its opponents so often said, the very opposite of enlightened morality. It seems to me that Protestantism and the Enlightenment were allies only when they faced common enemies. This, as we will shortly see, means that the relation must be stated in terms of periodization.

It is at least clear that Calvinism was not *always* clearly ranged against the Enlightenment. A generation of scholarship has shown us that the main citadels of New England Puritanism were the first to let in the Lockeian infiltrators. Much later, Presbyterian Princeton was the entry point for what many have called the Scottish Enlightenment. During the Revolution, Nature's God and Jehovah were clearly on the same side if never quite the same Person. Yet much of the settled Calvinist clergy was marshalled by Jedidiah Morse and others in a solid phalanx against French infidelity in the late 1790s.

Calvinism, then, was not always and everywhere the enemy of the Enlightenment. On the other hand, I think that popular revivalism was. Alan Heimert has argued the contrary with great learning, and it is clear that not all revivalists were uneducated or obscurantist.[7] Yet their universe was different from that of the Enlightenment; what was most important was what was most mysterious. Even Edwards, I think, made eager use of the weapons of the Enlightenment in a direction contrary to its general direction as I have loosely defined it.

Mass emotion, or what its eighteenth-century opponents called enthusiasm, has been, I think, the most powerful opponent of the Enlightenment in America. This is not by any means to say with Mencken that the educated few have always been right or that the excited have always been mistaken. Mass emotion has come sometimes from the right, sometimes from the left; it has occurred as often in political as in religious terms. But its style and method, its idea of what is true and what is most important are opposed to those of the Enlightenment.

Since this kind of opposition was not consistent but came in waves, the American Enlightenment was without the advantages of a solid, institutionalized enemy. This lack, I think, explains many of its characteristics. It has lacked such eighteenth-century characteristics of matter as resistance and solidity; to change the figure, it has been most successful not in attack but in permeating or being absorbed.

If one bears in mind this amorphous quality of the American Enlightenment, resulting in part from the nature of its opponents, it is a little easier to understand what ideas from Europe were most and least successful in crossing the ocean. In all fields, what flourished were ideas susceptible of compromise.

In philosophy, Americans accepted the ambiguities and contradictions of the Lockeian tradition, especially in its Scottish version, with even less difficulty than Europeans. The primacy of sensation and the centrality of the moral sense could flourish at once, as long as each of these theories behaved itself and did not push too far. Hutcheson and Clarke early, Reid and Stewart later, seemed to answer the questions Americans wanted to ask. Hume, though he was sometimes read, was usually rejected, and Kant ignored until much later.[8]

In science, we understand thanks to its historians that neither natural history nor natural philosophy seemed in this period to present really dangerous problems. If more Americans tended to study the former, it was partly because it demanded less mathematics and answered more urgent local needs. For reasons, apparently, both of social demand and intellectual bent Americans tended even more than Europeans toward the useful and the improving.

In religion the rising view in the middle or late eighteenth century was what Conrad Wright has so fruitfully described as supernatural rationalism: the belief that intuition, experience, and reason will all prove to complement and confirm, rather than displace, the Christian revelation.[9] Even conservatives who resisted supernatural rationalism were forced to enter its world; it may be guessed that only the simplest kinds of pietism remained completely unaffected by it.

Just what was essential in the revelation that reason was to supplement was a matter on which there was a good deal of disagreement. But even on the left of the enlightened spectrum the extremes were rare. Deism, not real scepticism, was the main threat seen even by the most worried of the orthodox. Material- ism seldom went farther than the odd kind Jefferson picked up from Priestley, who in turn seems merely to have rigidified the cautious suggestion of Locke that God *could* give matter the power to think. In America, perhaps more easily than in Europe, it was possible to believe in something called materialism and yet to hang onto moral certainty and even to leave some vague entity to be punished or rewarded after death.

No extremes seem to have flourished in America, either of sys- tem or of negation. Thoroughgoing materialism, Utopian radi- calism, or even systematic deism seem rare on the left, and so do cosmic optimism or esthetic pessimism on the right. Primitivism seems to have been half-hearted where it existed; the savage looked less noble when you saw him close at hand and wanted his land. Cynicism was rarest of all; there was no place in Amer- ica for anybody like an unbelieving abbé or a nobleman who pa- tronized sedition. Of Americans, only the Protean Franklin could occasionally sound as wicked as Voltaire, and he kept his most penetrating wit for occasional pieces, some of them privately cir- culated. A list of Europeans to whom there is nothing remotely resembling an American counterpart might start with Lord Ches- terfield or Dean Swift, and go on quickly to either Gibbon or Hume.

In politics, we know that the range is similarly limited. Enor- mously fertile and original as it was, American political thinking of the Enlightenment found its opposing archetypes in two won- derfully symbolic figures, separated only in crisis, friends in youth and old age. Real conservatives in religion and politics, like Jo- seph Dennie, were not very common, and real radicals in both like Joel Barlow (in his later career) still rarer. Nothing could be less useful than to try to associate philosophical, religious, and political views in any rigid manner. In thought as in political organization, Madison's extensive republic was saved from faction by its complexity.

It is perhaps significant that when historians have looked beyond the familiar and fascinating group of statesmen for representative men of the American Enlightenment, they have looked most often at college presidents. Then as now, there were solid reasons why these were seldom spokesmen of very advanced or consistent views. Some, like Dwight, were extreme opponents of the radical Enlightenment in the 1790s. Before and after that crisis they were more likely to be conciliatory liberals or moderate conservatives.

How widely, to proceed to our next question, did even the moderate Enlightenment penetrate in the American population? My own lack of knowledge, the lack of any possible clear data, and the lack of rigid structure in American society itself all enjoin an especially cautious answer. And the question of spread cannot be discussed at all except in conjunction with the next question: that of periodization.

Provided one's definition of the American Enlightenment is sufficiently flexible, I think that one's periodization can be rather precise. The history of the American Enlightenment can be divided into three periods. The first, running from the arrival of the New Learning at Yale in 1714 until the end of the Revolution, is a period of quiet growth and spread. The second, starting in 1784 with Ethan Allen's *Reason the Only Oracle of Man,* is one of radical offensive and conservative counterattack. The third, beginning in 1800 and essentially over by about 1815, is the period of defeat and absorption.

To return now to the question of spread, in the first period the ideas of the moderate English Enlightenment affected most educated Americans. Locke and the popularizers of Newton became standard fare in the colleges. The Virginia gentry, though most were by no means Jeffersons or Madisons, expressed sceptical ideas in their private correspondence as if these were the common property of sophisticated people. Doctors and merchants dabbled in natural philosophy. We do not really know—or certainly I have little idea—how many apprentices and small merchants were as lively-minded as Franklin's early friends. Perhaps it *may* prove true that enlightened ideas penetrated farther down in the less rigid and more amorphous social order than they did in Europe. I think, however, that they remained within the confines of the urban middle class except in the South.

In this period before the Revolution there was less difference than later between the American and European Enlightenments. The peak of French radicalism only began about 1750, and its writers did not reach North America fast. Most of the Enlightenment books that turn up in American libraries in this period are of the mild English or Scottish variety: Locke, Clarke, Hutcheson. Yet a little more subversive stuff was slipping through the mental barriers: Bayle's *Dictionary* and Hume's essays. Voltaire's *English Letters,* a common item on library lists along with his histories, contains plenty of carefully disguised dynamite.

Before the Revolution, however, there was no decisive confrontation between the Enlightenment and its enemies. The Great Awakening, like Wesleyanism in England or, for that matter, many kinds of enthusiasm in France, served to draw some distinctions between the partisans of emotion and the defenders of intellect. It served perhaps to circumscribe, even to localize, the Enlightenment. But the controversies had to some extent died down by the Revolution, and it proved possible for New England Calvinists and Virginia rationalists to work together without too much difficulty. This was made clear by the fact that the school of "rational dissenters" in England, among them Richard Price and Joseph Priestley, were among the most useful allies of the colonists.[10]

The second period, that of militance and counter-militance, was I think partly a result of post-revolutionary malaise. Whether or not American political institutions worked badly in what used to be called the Critical Period, the first revolution of modern times was thrilling to some people and frightening to others. The new flowering of republican literature and science seemed slow to arrive. High hopes were balanced by a nearly disastrous disruption of institutions in some states, and by a breach in the formerly easy flow of European ideas. Some kinds of upper-class Enlightened people left, others lost authority.

Thus all sorts of hopes and fears for the new republican order made partisans and enemies for the new and more radical French Enlightenment. It took a decade for the conflict to reach its full intensity. Ethan Allen seemed to link deism and popular sovereignty in 1784, and Timothy Dwight assumed the role of defending religious and social order with his *Triumph of Infidelity* in 1788. Yet well beyond this date, and even beyond the Revolution and Reign of Terror in France, as Mr. Gary Nash has shown, many of the New England clergy were finding in the downfall of king and church only further evidence of God's judgment on corrupt political and religious institutions. Only in 1795, while the Thermidorians were defending property and punishing revolutionaries, did the tocsin really sound from a considerable number of New England pulpits.[11]

The reasons for this curiously belated panic are, I think, many. Some lie in the new aggressiveness of the American Deists, with their bold appeal to the public through their clubs and newspapers. The most important single event was the timely publication of the *Age of Reason*, with its appealing simplicities and wide distribution. The suggestion that political revolution should obviously be followed by social and religious change put conservative revolutionaries on the defensive. Some of the most alarming European books were published in America for the first time in 1795, including Holbach's *Christianity Unveiled* and Voltaire's *Philosophical Dictionary*, in which he boldly showed his anti-Christian colors. Yet the causes of alarm extend beyond the militance of the radicals. Some are to be found, I think, in growing religious pluralism and threats from within to the Standing Order in New England, others in fears for the continuing ascendancy of the party of Washington and Adams. All these dangers and doubtless others, some of them perhaps subtler and deeper, seem to have come together about 1795. Something like this combination was necessary before any sizable number of Americans could thoroughly surrender to their latent fears. Until then some had believed American institutions too radical; few had seen them as seriously threatened in their turn from the left.

For whatever reason, we all know that the period from 1798 to 1801 saw hysteria rampant and that then, for the first time, religious doubt and political radicalism were widely linked to what one best-selling tract called the Cannibal's Progress. The picture became familiar and soon monotonous: the Bavarian Illuminati had from the beginning been the promoters of world revolution, the secret paymasters of Voltaire, Robespierre, Paine, and Jefferson. As always the monolithic character of the two sides can be exaggerated. Some Calvinists and many pietists and evangelicals stuck to republicanism, or Jefferson could never have won. And a good many Federalist gentlemen undoubtedly took the threat of infidelity a little less seriously in private than in public. Yet in fairly widespread circles, the Enlightenment was for the first time seen as a party, rather than a universal tendency of history.

One of the dangers for those who want to assess the power of the radical Enlightenment in this period is to take the evidence uncritically from its opponents. To make use of Lyman Beecher's remark about deism at Yale, reported by his children as part of his reminiscences a generation later, is even worse than it would be to make Senator Joseph McCarthy the chief source for the spread of Communism at Harvard in the thirties.[12]

Where there is smoke, as somebody said of McCarthyism, there is a smoke machine. Yet the hearings of congressional committees may have made more Americans aware of the existence of Marxism than the devoted efforts of Marxist propagandists. And in a somewhat similar way, the literature of panic and alarm of the 1790s may have informed a good many Americans about the godless French. Some of their clerical enemies, like Nathanael Emmons, plowed conscientiously through their works. Voltaire and Hume became, perhaps, more familiar than ever before at least as names.

Probably never very strong, the attack of the radical Enlightenment was turned back, not without damage to the moderate Enlightenment as well. Revivalism was the most successful weapon of the opposition. With the Church of England nearly destroyed it had lost one important opponent and others had been weakened. In the South, particularly, evangelical piety almost destroyed the aristocratic, usually Anglican Enlightenment; sceptics or scoffers died off, were converted like John Randolph, or decided they had best shut up. On the academic front, European allies were called in; among the common books in American libraries were those of Joseph Butler, William Paley, and even Edmund Burke.

The most conservative and one of the oldest wings of the American Enlightenment lived on with little alteration in Boston Unitarianism. The radical Enlightenment of the late 1790s dwindled to a vestige. One mid-century memoir after another recalls with amazement how Voltaire and Paine were widely read, by quite respectable people, in the authors' youth. In American villages, the dangerous thinkers of 1850 and later were to shock their neighbors with opinions that had been tame in Paris a century before.

The most important and most widely received ideas of the American Enlightenment, however, were not eliminated but assimilated. No process is more important for the understanding of American culture than the way potential enemies are taken over. Not for the last time, denatured radicalism became part of the mild progressive mixture.

To many people's surprise, it became quite clear that republicanism, even political democracy, was compatible with property rights and could even concede a special role to a cultural élite. In philosophy radical rationalism and radical empiricism were further tamed by the reassuring formulae of common-sense philosophy. In religion the great doubts and fears were once more assuaged—for many people—by reiterated demonstration that Revelation was confirmed by the book of nature and the natural inclinations of the mind. Above all, morality in the arts, in public life, and in private conduct was demonstrated with endless reiteration to be the necessary principle of republics. As for the eighteenth century, it was soberly summed up by the Reverend Samuel Miller of Princeton and others as a period of great progress in natural philosophy, useful invention, republican government, religious freedom, and polite letters. All this could be granted while pointing out that its record in morals and religion was far more dubious.[13]

One can perhaps summarize this periodization without stretching things too much by saying that in the first period the American Enlightenment drew chiefly from England, in the second from France, in the third from Scotland. If the period of gradual spread was characterized by supernatural rationalism, and the period of militance by deism, the period of absorption was given its central meaning by moralism: the belief that accepted moral dogmas are part of the framework of the universe.

Among recent judgments on the American Enlightenment, perhaps the least flattering is that of Peter Gay, who has compared it in importance to the Scottish or Genevan rather than to the English or French or German Enlightenments.[14] If one's criterion is the creation of original philosophic systems, this may even be generous. Yet if one is talking about political institutions or social ways of life, the American Enlightenment is as important as American nineteenth-century culture, whose way it prepared.

To talk about importance is to sidestep value judgment, and this historians, however tentative their research, should try not to do. In my opinion the Enlightenment everywhere was an unstable mixture. In Europe the epistemological paradoxes which lay at its center led in many desperate directions: to rigid materialism or positivism or utilitarianism, to totalitarianism, to intellectual or moral nihilism, or by reaction into the logical madnesses of extreme romanticism.

America, until comparatively recently, managed to avoid these and other damaging extremes. And even in 1969, when compromise and consensus are so unfashionable, their value must not be lightly dismissed. The American political experiment was a bold one, and it badly needed some moral cement to make construction possible. It remains true, as Adrienne Koch has pointed out, that the American Revolution did not, like other modern revolutions, devour its children.[15]

Yet for all the hard work that went into it and the achievements it has made possible, one cannot today see the American compromise without some ambivalence. From the common-sense philosophy and criticism which ruled polite America in 1815 came much of the defensive quality that gave American literary culture for so long its genteel and slightly archaic tone. The habit of blunting contradictions has sometimes prevented Americans from seeing contradictions that are real, especially the one tragic exception in America to even the minimum of Enlightenment freedom and equality. Ambivalent about change, given to accepting past enlightenment as satisfactory, American thought has been a baffling mixture of conservatism and innovation. Nowhere have the advantages and disadvantages of avoiding intellectual confrontation been more apparent than in the history of American religion, where popular success and even humanitarian achievement have often been purchased by surrendering of meaning.

American academics have been forced by recent events to choose sides on many questions. I do not think that we need exactly to approve or condemn the American Enlightenment or its opponents, to fight again the battles of the 1790s. The compromises formed by about 1815 served us well, or seemed to, for a long time. Inescapably, however, they, like the whole of our past, will look different when we emerge from the cultural revolution through which we are now passing.

Notes

1. Paul Hazard, *The European Mind, 1680-1715,* tr. J. Lewis May (Paris, 1935; paperback ed., Cleveland, 1963).
2. Peter Gay, *The Enlightenment, An Interpretation* (New York, 1966).
3. Ernst Cassirer, *The Philosophy of the Enlightenment,* tr. Fritz C. A. Koelln and James P. Pettegrove (Tübingen, 1932; paperback ed., Boston, 1955). Leslie Stephen, *English Thought in the Eighteenth Century* (London, 1876; paperback ed., 2 vols., New York, 1962).
4. Lester J. Cappon, ed., *The Adams-Jefferson Letters,* 2 vols. (Chapel Hill, 1959), pp. 493-94, 496, 501, 510, 515.
5. Boorstin's most extreme statement of this thesis is his essay, "The Myth of an American Enlightenment," in *America and the Image of Europe* (New York, 1960), pp. 66-78. But see also, for a more general argument to the same effect, the immediately preceding essay, "The Place of Thought in American Life," *ibid.,* pp. 43-61, and, for strictures against the application of European categories to the history of American culture, *The Americans, The Colonial Experience* (New York, 1958), pp. 393-94; and on shortcomings of intellectual historians, *The Americans, The National Experience* (New York, 1965), pp. 436-37. Boorstin's view is effectively criticized by Bernard Bailyn in his discussion of the relation between the Enlightenment and the Revolution in "Political Experience and Enlightenment Ideas in Eighteenth-Century America," *American Historical Review,* LXVIII (1962), 339-51.
6. Carl Bridenbaugh, *Mitre and Sceptre* (New York, 1962; paperback ed., New York, 1967).
7. Alan Heimert, *Religion and the American Mind* (Cambridge, Mass., 1966).
8. Statements in this article about early American reading of European books are based on tabulations of libraries, book sales, auctions, and American editions. [For those tabulations, see David Lundberg and Henry F. May, "The Enlightened Reader in America," *American Quarterly,* XXVIII, 2 (Summer 1971), 201-11 plus graphs.]
9. Conrad Wright, *The Beginnings of Unitarianism in America* (Boston, 1954).
10. See Anthony Lincoln, *Some Political and Social Ideas of English Dissent, 1763-1800* (Cambridge, England, 1938).

11. Gary B. Nash, "The American Clergy and the French Revolution," *William and Mary Quarterly*, XXII (1965-66), 392-412. It is perhaps worth reporting that shortly before the publication of this article I had reached exactly the same chronological conclusion from a less systematic examination of sermons.

12. Beecher, who was at Yale from 1793 to 1798, recalled that "That was the day of the infidelity of the Tom Paine school . . . most of the class before me were infidels, and called each other Voltaire, Rousseau, D'Alembert, etc. etc." *Autobiography*, ed. Barbara M. Cross, 2 vols. (Cambridge, Mass., 1961), I, 27.

13. Samuel Miller, *Retrospect of the Eighteenth Century*, 2 vols. (New York, 1803). Many similar assessments are available in the century-sermons preached at the close of 1799.

14. Gay, "The Enlightenment," statement read at a meeting of the American Historical Association in December, 1966.

15. Koch, *The American Enlightenment* (New York, 1965), p. 19.

· 7 ·

The Decline of Providence?

The title of this paper was assigned when I was asked to speak at a plenary session of the 1975 International Conference on the Enlightenment, held at Yale. These conferences have a large French component, and I prepared the paper with a French audience in mind. However the attention of my audience, which was indeed heavily French, was distracted by a closer-to-home argument.

The paper was paired with one by Professor Georges Gusdorf, of Strasbourg, on the same topic. The paper was excellent, but Professor Gusdorf added a specific attack on what he saw as the anti-religious and especially anti-Protestant prejudices of the French academic establishment. This gave rise to an uproar, in which Gusdorf was accused of being a traitor to the tradition of the *philosophes*. I found that passionate reaction instructive. I have since learned in other international meetings that some intellectual questions debated in the eighteenth century still have highly emotional resonance in Europe, and especially in France.

I hope that somebody, some day, will develop my tentative classification of deisms. An international social and intellectual history of deism, treating it seriously as a religion, might make an excellent book.

This paper, like all plenary-session papers for these conferences, was published in *Studies on Voltaire and the Eighteenth Century*, CLI-CLV (1976), 1401-16, and is reprinted by permission of the editors.

I find myself in general agreement with Professor Gusdorf, although my research deals with another continent and has a somewhat different emphasis. Whether this agreement is providential, and whether Providence in this case has made use of the organizers of this conference, I will not attempt to determine. In the course of my paper some minor disagreements with my colleague will become clear. I support cordially however his major assertion, that the eighteenth century saw a series of religious changes rather than a decline of religion, and I think that this fact is still clearer regarding America than Europe. I agree even more cordially with Professor Gusdorf's statement that one cannot study the movement of ideas of the eighteenth century—I would add of any other—without thinking hard about the history of religion. In America more often than in France, I think, recent historians of many kinds and many beliefs have been giving closer and more expert attention than they used to to religious studies.[1] It is true in America as well as France, however, that the Enlightenment has not been sufficiently treated in terms of religious history, and that this has led to some misunderstandings. In my opinion Rousseau and Paine and Jefferson were almost as religious in temperament as Wesley and Edwards. They were deeply concerned about the nature of the universe and man's place in it. And their ideas on these subjects were formed by the Christianity which they partly or wholly rejected.

A neglect of religion or of theology is one common error among historians of the Enlightenment. Since this point has been made by Professor Gusdorf, I should like to begin my paper by mentioning three other faults, common among writers of intellectual history in general, and particularly among those who deal with the Enlightenment. First, despite much criticism on this point, historians tend to talk about the movement of ideas in too general and sweeping a manner. (Here I am not criticizing Professor Gusdorf—in a short paper on a subject like this one must necessarily be sweeping and general, as I too am about to be.) If, in any extended treatment of the Enlightenment, one says that Newtonian ideas spread, or scepticism increased, or religion declined, one must say when, where, and among whom. Obviously there are great differences between Protestant and Catholic Europe, in America among North and South, East and West, between city and country everywhere, among classes and occupations. Intellectual history need not give place to social history—provided it works in close alliance with it.

In connection with the Enlightenment it is important, I think, to remember that those who found it possible to believe in an orderly and regular universe were usually those who were able to live orderly and regular lives. People who were subject to the vagaries of weather, the ravages of war or revolution, the upheavals of economic change, the crushing impact of epidemic or famine, were seldom tempted to agree with Pope's unguarded statement that "whatever is, is right." In the eighteenth century, historians often tell us, life was growing more secure. They should always say rather: more secure for certain groups in certain countries, and here intellectual historians must learn what they can from the voluminous, scattered, and as yet very incomplete researches of the social historians.

Of those who were not secure, many found it necessary and helpful to attribute their sufferings to the will of an incomprehensible and mighty God, who tries and chastens those he loves best. Some placed their hopes in a future when those who have suffered for righteousness' sake will be rewarded and their persecutors will perish. This millennial future could be secular and revolutionary rather than biblical. In America and England at least, it could be both. Among many others, a number of the New England clergy interpreted both the American and the French Revolution, with somewhat different emphases, in terms of the working out of the divine plan and the imminent approach of the end of the world.

My second criticism of historians of the movement of ideas is that they sometimes concentrate so much on doctrine and credo that they fail to take due note of image and vocabulary, of feeling and tone. There are ways in which Rousseau, Jonathan Edwards, Wesley, Wordsworth, Godwin, and Blake all have something in common as against Pope, Hume, Voltaire, Archbishop Tillotson, Samuel Clarke, Benjamin Franklin, and James Madison. Obviously the members of these two groups differed among themselves on almost every doctrinal point, and the grouping can be used only for certain special purposes related to feeling and style.

This brings me to my third general point about intellectual historians. They must avoid reifying and conventionalizing their categories. Though their main business may be with general movements of thought, they must always remember that such movements are made up of individuals. Individuals are never fully defined by any category, and must often be discussed in one category for one purpose and in another category for other purposes. To put this in a different way, the history of ideas must learn from social science, but it must remain fundamentally a humanistic discipline.

These general precepts are not at all new, and I believe most people who work in intellectual history accept them. Yet they are sometimes forgotten by all of us. If intellectual historians observed them more carefully, in dealing with the Enlightenment as with other subjects, we would hear less discussion about whether intellectual history is any longer a legitimate pursuit.

I should like now to examine our topic mainly in terms of the United States, which is the country I know most about. I hope particularly that what I have to say will prove interesting to historians of Europe, European and American. In terms of the history of European ideas, I agree with Peter Gay that the American example is not of overwhelming importance. Most of the influence of America on the European Enlightenment was an influence not of an actual America but of the America that some Europeans wanted to believe in, an America that has for centuries disappointed its European admirers.

Yet I think European historians might well find it interesting and profitable to study the familiar Enlightenment ideas refracted by a different society: a society largely agrarian, with a vigorous bourgeoisie but no great cities; a society profoundly Protestant but already diverse both in national and sectarian origin, a society quasi-republican in politics well before the American Revolution; a society where every idea was affected by the presence of a frontier at once promising and menacing, and also by the presence of three races. America was not Europe, and yet part of the interest of this period of American history comes from the fact that some Americans were trying hard to be European. European books, including nearly all the major classics of the Enlightenment, circulated in America quickly and widely. They were, of course, read selectively. Yet it seems probable that in the late eighteenth century, at the same time when Americans began the long process of separation from Europe, the American colonies of England were more European than they had ever been before, or than the United States was ever to be in later times.

As in Europe, much damage has been done to the understanding of the Enlightenment in America by the two false pictures coming from right and left, whose damages Professor Gusdorf has clearly indicated. In 1798 certain prominent members of the New England Calvinist clergy, deeply disturbed by religious, social, and political change, came to the conclusion that the whole Enlightenment had been a conspiracy, stemming from the Illuminati and including Voltaire, Hume, Rousseau, Paine, and Jefferson—a group some of whose members detested the others. Jedidiah Morse and others avidly circulated the works of Europeans who said this: especially the Abbé Barruel and Professor John Robison of Edinburgh. In less lurid colors, this version of the history of the Enlightenment was repeated through the nineteenth century, especially by clerical historians, who recited the exciting legend of the near triumph of "French" atheism, and its defeat by something like a popular Christian crusade.

We may think that the conspiratorial version of the history of the Enlightenment troubles us no longer. I learned with interest that Robison's *Proofs of a Conspiracy* was republished in 1967 by a right-wing publisher as one of the "Americanist classics." In this edition an introduction adds the names of Galbraith, Rostow, and Lippmann among others to the conspiracy of enlightened "half-baked intellectuals" who run America[2].

Most of the power of this homogenized version of the Enlightenment however comes as in Europe not from the right but from the left. In America it comes mostly not from the far left but from the much more powerful moderate and liberal left to which most of us academics usually give our allegiance. This version, I think, reached its greatest strength in America in the late nineteenth and early twentieth century among pragmatic social scientists, among them James Harvey Robinson and, in far less naïve form, John Dewey. These men, like many others in all periods and especially in the eighteenth century, tended to think that all history inevitably led to themselves. Thus, as they saw it, the "mind of the eighteenth century" moved gradually from orthodoxy to natural religion to deism and and finally to pragmatic secularism. Doubtless for some individuals or groups of individuals this scheme is defensible. Stated as a generalization about the movement of thought, however, it fails to fit American actualities. Its main deficiencies are two: it makes it impossible to explain American nineteenth-century culture, and it leaves out most of the people in all periods.

The progressive and secularist view of American history in its

early twentieth-century form has come under attack from many
sources. Its optimism has been attacked from the right and—at
present more powerfully—from the left, which tends to object
to any cheerful treatment of the course of American history.
More than in Europe, its secular emphasis has also been called
into question by the increasingly numerous and well-informed
historians of religion.

Perhaps it is time to suggest a substitute for the general thesis
of the decline of religion throughout American history—a thesis
that cannot be sustained and is indeed refuted by the best
statistics available. One might suggest that in most periods, real
scepticism has been confined to small groups of intellectuals.
Religious liberalism has been characteristic mainly of the upper
middle class. Bland religious lip-service has been general among
the middle class as a whole: here as in Europe it is easier to
measure church attendance than religious intensity. On the other
hand, emotionally fervent and theologically conservative religion has always been very powerful, especially among the poor
and oppressed, and is gaining fast at present. Whether or not
these impressionistic statements will hold up completely, I think
that they are more nearly correct than simple statements—still
often made—about religious decline.

I should like now to abandon generalization about intellectual
history and about America in all periods, and return to our
specific topic. I will try to summarize some of the principal
kinds of belief in Providence which prevailed in America in the
eighteenth century. I shall do this first without regard to numbers or chronology, and then very briefly try to make some
statements about both of these. I will deal first with Christian,
and then with non-Christian views.

Within a rich variety, most eighteenth-century Americans
were one kind or another of Calvinists, and the most powerful
churches subscribed to the Westminster Confession, whose doctrine of Providence is clear: "God the great creator of all things,
doth uphold, direct, dispose, and govern all creatures, actions

and things from the greatest even to the least by his most wise and holy Providence, according to his infallible fore-knowledge, and the free and immutable counsel of his own will, to the praise of the glory of his wisdom, power, justice, goodness and mercy."

In bringing about his great objective, the manifestation of his own Glory, "God in his ordinary Providence maketh use of means, yet is free to work without, above, and against them [that is, miraculously] at his pleasure." His Providence includes the Fall, but man is solely responsible for sin[3]. In working out the contemporary meaning of these doctrines, New England preachers still found it perfectly clear that God often made use of disasters and of the energies of the wicked.

A millennial view of the operations of Providence was common among Calvinists and other Protestants, both in America and England (here I may differ from Professor Gusdorf). To students of prophecy, all events and especially the most terrible have an ultimately benign meaning. Thus for many eighteenth-century Anglo-Americans all their troubles—Indian massacres, French triumphs on the frontier, British oppression, the triumph of Revolution and the destruction of Catholicism in France—all these were ways of bringing on the reign of peace and love, which could not be far off. From time to time, American theologians saw a special role in this ultimate set of events for that sorely tried but also cherished portion of the universe, America.

A special view of Providence was that of the ultra-Calvinist intellectuals stemming from Jonathan Edwards and exerting considerable influence in the second half of the century. This view, Augustinian and in the view of its opponents verging on pantheism, held that constant active exertion of the will of God was necessary to maintain the universe in existence and to maintain any relation between actuality and the content of our minds. Among some of the later Edwardsians, men who were willing to bite the hardest bullets, God becomes for his own purposes the actual author of sin. For some of these men, as for some European quietists, people are required to acquiesce joyfully in their own damnation, if this be necessary for the ultimate end of the universe, the glory of God. Such extreme views could hardly be the basis of truly popular religion, but their spokesmen controlled many pulpits and seminary chairs.

At the opposite extreme among Christians were the liberals and Arminians. These drew their views from Locke, from the Anglican latitudinarians, and from liberal dissenters in England, all of whom were very widely read in the American colonies. Some liberals were to be found in the Church of England in America. The single most cohesive and articulate group was in the Bostonian wing of the Congregational Church, which contained a number of especially eloquent liberal preachers.

The central doctrine of liberal Christians was that Calvinism was wrong about the nature of Providence. God could not decree eternal punishment because this would be useless and immoral. His nature was such that he could not act in an irrational manner; he was bound by his own natural laws. He had moreover given us faculties sufficient to recognize truth and make moral judgments. The necessary moral truths could be read by all in the book of nature, which revelation supplemented but could not possibly contradict. From this start stemmed a tendency toward progressive rejection of all paradox and mystery: by the middle of the century advanced thinkers in Boston believed, though they seldom quite said, that all men must eventually be saved. They had begun to question both the atonement and the Trinity, though they stopped short of deism and insisted that their views were consistent with a correct reading of Scripture. To these liberals, the cheerful and rational truths in which they believed were being ever more clearly understood among educated people. Yet their spread was hampered by the preachings of ignorant enthusiasts, who appealed to the passions of the vulgar.

Most Americans throughout the century were neither extreme theological Calvinists nor liberal Christians but rather Christian pietists of generally Calvinistic inclinations. In the Great Awakening of the mid-century in particular, the most successful revival preachers usually urged their congregations to return to the doctrines of grace rather than works. But they also often taught that these doctrines could be best understood by the heart and not the head. For the multitudes of simple pietists, God's intentions were not to be precisely predicted. Their generally reassuring nature was made clear by the current revivals of religion themselves, which were clearly part of a great millennial plan.

Among the minority of non-Christians there was also a wide variety of ideas and feelings about Providence and all other matters. Professor Gusdorf has pointed out that in Europe the word "deist" could cover many kinds of people; so could it in America. In a brief attempt at classification one might start with complacent deism. Among themselves cultivated planters or merchants might easily concede that miracles and revelation were necessary for maintaining the morality of the common people. For those able to read the classics, however, these props were not necessary, and the narrative of the Bible was clearly not to be taken seriously as a description of the way God acted. Enlightened men believed rather in a God who acted much as they would if they had been in his place, moderately, sensibly, and with clear meaning. This view led directly and inevitably to Pope's conclusion or that of Doctor Pangloss, that all is for the best.

Some deists were too sensitive and intelligent to follow this complacent path, so the next category might be called agnostic deists. These are people who simply gave up trying to understand the way of Providence, or to deal intellectually with the problem of evil. Among them we can clearly place Benjamin Franklin, who found a benign agnosticism increasingly congenial in his cheerful old age. John Adams, who tried hard to be a Christian, was sometimes driven to take refuge in a position close to that of the agnostic deists: "Mr. Adams leaves to Homer and Virgil, to Tacitus and Quintilian, Mahomet and Calvin, to

Edwards and Priestley, or if you will, to Milton's angels reasoning high in pandemonium, all their acute speculations about fate, destiny, fore-knowledge, absolute necessity, and predestination. He thinks it problematic whether there is, or ever will be, more than one Being capable of understanding this vast subject."[4]

Another group might be called troubled deists. These were men who made a sincere and devout effort, not always successful, to find a God less cruel and paradoxical than the God of the Bible. Among these was Thomas Jefferson, who needed deeply to believe in the beneficence and rationality of the Creator, subscribed formally to this belief at all times, but suffered too much to be able easily to maintain it.

(The categories of course overlap.) A belief in Providence pervades not only Jefferson's moral and political theory but also his views of the cosmos and the biological world. This was true of the other enlightened scientists of his Philadelphia circle. We can thus call another category scientific deists. One of the boldest of the Jeffersonian circle was Benjamin Smith Barton. In discussing the devoted efforts of birds to defend their young, Barton found that this demonstrated "in language most emphatick, the existence, the superintendance, the benevolence, of a first great cause." Barton was, he said, "always happy," in the study of natural history, "to discover new instances of the wisdom of providence. . . ."[5]

A final category can be called radical deists. These saw the chief evidence of God's intentions in the movement of history. Some Americans toward the end of the century came to believe, like some Frenchmen and Englishmen, that a great secular and republican millennium was beginning. In the near future superstition, fear, tyranny, greed, persecution and war would disappear together. Among those who for a time believed deeply in this future were the important poets Philip Freneau and Joel Barlow, and the chief public preacher of deism, Elihu Palmer. All three of these men, perhaps significantly, had been trained in Calvinism in their youth. In 1793-94 the least sceptical of deists, Thomas Paine, devoted his great polemical talents to proving that God was too wise and good to have anything to do with the Bible. Nobody believed more deeply than radical deists in an all-wise Providence. Their mood was very similar to that of the simpler kind of revivalist Christians. Nothing could be more different from the mood of earlier, complacent and aristocratic deists, though the theological content in all kinds of deism was much the same.

So far my list of attitudes toward Providence has been purely taxonomic; it has taken no account of time or space or numbers. In a short paper I cannot deal with the important matter of regional variation. Numerical statements in these matters still have to be made with the utmost caution; I am trying to make a few elsewhere[6]. I should like however to sketch very briefly the chronological development, through the century, of these various views and even of their popular strength.

The first half of the century probably saw some decline of religion though one must beware of later pious exaggerations, made for the purpose of increasing the importance of revival. Such decline as there was was a matter of neglect more than of the penetration of English deism, which reached only a very few. Prosperity and complacency in the cities was a part of it, and so were frontier ignorance and sectarian bickering.

This period was brought to an end by the Great Awakening of the 1740s and '50s, mostly amounting to a revival of Calvinism. The widespread and sometimes dramatic revivals of religion were expected by their leaders to restore belief in the doctrines of grace rather than works, and by some of them to aid in bringing on the millennium. While the Great Awakening did strengthen American Christianity, and especially its Calvinist section, it left the churches contentious and divided. Moreover, in something like dialectical opposition to the Awakening, liberal Christians made their first major and specific gains. Everywhere refugees from harsh doctrines and bad manners flocked into the Anglican church, the church of moderation, polish, order, and up-to-date ideas. In the Boston region, Arminians, still within the established Congregational Church, mounted a polemic against ignorant extremism, sturdily insisting that the Chrisian religion was always and everywhere rational.

As the tide of revival waned in the 1750s and 1760s, religious polemic partly gave way to political argument[7]. The period of the Revolutionary struggle, let us say from 1763 to 1787, can be seen as a time of religious truce. Calvinists, Arminians, and deists all took part in the resistance to what was seen as British tyranny. The two superb propaganda documents of the Revolution show the felt need for religious compromise. Who could reject Jefferson's appeal in the Declaration to the 'laws of nature and of nature's God'? To make the document still less objectionable on religious grounds, the Congress added to Jefferson's draft the phrase 'with a firm reliance on the protection of divine Providence'.[8] Thomas Paine, in his great pamphlet "Common Sense," said nothing to offend Christian or Calvinist sensibilities. Government, he declared, was the badge of lost innocence—this could surely be read by those so inclined as a reference to the fall of man. Paine contrasted the "royal brute" of Great Britain to the true King of America, who reigns above[9]. Small wonder "Common Sense" was read in pulpits as well as camps.

In the revolutionary coalition of people of different religious beliefs, Calvinists were conspicuous for their patriotic fervor and unity. Why not: in their own eyes they were defending their ancient civil and religious liberties against decadence, corruption, and luxury; against tyrannical officials and worse, the specter of bishops.

This brings us to a fact which seems to me of the first importance for American history and for understanding the difference between America and Europe. In America, unlike either England or France, the dominant form of Christian orthodoxy was clearly ranged not on the side of ancient oppression or the defense of the status quo, but on the side of successful revolution.

In America, Providence, and specifically Christian and Calvinist Providence, was on the side of the people.

Religious truce continued through the period of the Constitution, which was made by an interesting coalition of moderate Calvinists, moderate deists, and Episcopalians, who are moderate by definition. This was a period of comparative religious calm, and perhaps partly in the interest of unity the Constitution, unlike the Declaration of Independence, omits all mention of Providence—much to the disquiet of later pietists. Shortly after the beginning of the new government, sharp political divisions began to form. The argument started over the nature of the new republic, and was given a special religious meaning by the dispute over the French Revolution.

At first American opinion was united in favor of the Revolution in France. Calvinists saw it in providential terms. At last their ancient Catholic and absolutist enemy, whose recent alliance had been an embarrassment, was being purged. France would emerge from her ordeal with a purer religion and a popular government. There was nothing unduly distressing to Calvinists about the fact that this was being brought about through unworthy instruments, and through the shedding of blood. Radical deists were still more cheerful about French events. It was in this period almost entirely that the radical and millennial kind of deism, the deism of Palmer and Paine, seemed to be gaining some strength. It was even then confined to small groups, but these were highly articulate.

The apparent growth of radical deism in America, and also the spread of the French armies into the sacred Protestant territories of Geneva and Holland, made it difficult for the orthodox by about 1795 to see the French Revolution in cheerful terms. It was only in the last years of the century that some of the orthodox Protestant clergy decided to mount a violent and organized polemic against the French Revolution and its friends in America, and also against its ultimate authors. Oddly, these included not only Rousseau and Paine but Voltaire and even the anti-enthusiast and anti-revolutionary David Hume. Though alarmed orthodoxy still believed in the coming of the millennium, it now seemed clear that before that happy day great trials were in store for the American republic, and that Providence needed the immediate assistance of its official supporters.

The resolution of American political and religious conflict that took place in 1800 was a very curious one. The political victory went to the party alleged to be on the side of France, and to its head, Thomas Jefferson, correctly accused of being a deist. This did not mean, however, that deism had defeated Calvinist orthodoxy. Both of these antagonists lost the battle for the nineteenth-century future. The election of 1800 was accompanied by a second profound and sweeping religious revival, which quickly spread beyond the control of orthodoxy and even of Calvinism. People who voted for Jefferson, especially in the South and West, especially Baptists and Methodists, saw him not as a prophet of deism but as the defender of free and pure Christianity against New England theocrats.

Thus at the beginning of the nineteenth century, American society was dominated by two great movements: fast growing political and social egalitarianism, and fast spreading and ever warmer religious revival. Eighteenth-century varieties of deism and scepticism nearly disappeared, to be revived later in nineteenth-century America only by small groups and scattered individuals. A loose coalition of Protestant churches prevailed. More or less willingly, the clergy surrendered the remnants of official support but successfully defended the right of moral supervision. Thus a new kind of increasingly untheological Protestantism was widely identified with the main currents of the age: democracy and nationalism, isolation and expansion. Pious historians demonstrated the beneficence of Providence by pointing out how America had been saved from British tyranny and French infidelity, and indeed from European contamination. The design of Providence seemed increasingly clear, and it included the spread to the whole world of free government, pure morals, and evangelical Protestantism. In some quarters, for a long time, Providence and America, Protestantism and democracy became almost indistinguishable.

Now let us return very briefly to our main question: Did Providence decline in the eighteenth-century world? I think Professor Gusdorf and I agree that a belief in Providence in various forms continued to dominate European and American thought. Indeed, perhaps if we extend its definition widely enough, some conception of Providence nearly always prevails in all periods, at least among people of European origin. Those who think history completely unpredictable, or those who really believe, in their hearts as well as their heads, that the universe is made up of atoms moving in a random fashion, have always been rather few. One can even say that some specifically Christian idea of Providence continued at the end of the eighteenth century to dominate the minds of most Europeans and Americans. It is clear that in a number of groups in several countries piety and evangelical zeal were growing warmer.

Yet systems of ideas, and also religious feelings, are affected by their enemies, even if their enemies are overcome. Perhaps one might summarize the movement of American religious history, for our present purposes, somewhat like this. In eighteenth-century America, a liberal or Enlightened idea of Providence faced the biblical idea of Providence in a series of confrontations. The outcome was a compromise in which both the Enlightenment and Protestantism were blurred and mingled, and out of this came the nineteenth-century culture of the United States.

Notes

1. I called attention to this in an article, "The Recovery of American Religious History," *American Historical Review*, LXX (1964), 79-92.
    Since that article was published, this "recovery" has continued, along somewhat different lines than those I then perceived.

2. Robison, *Proofs of a Conspiracy* (Americanist Classics edition, Boston and Los Angeles, 1967), Introduction (pages not numbered).
3. Williston Walker, *The Creeds and Platforms of Congregationalism* (Philadelphia, 1969), 372-73.
4. John Adams to John Taylor, 1814, reprinted in Adrienne Koch, ed., *The American Enlightenment* (New York, 1965), 222.
5. Barton, "A Memoire concerning the fascinating faculty which has been ascribed to the rattle-snake, etc.," American Philosophical Society, *Transactions,* IV (1799), 107, 108.
6. An article discussing the relative popularity of various European authors in America, written in collaboration with David Lundberg, will appear in the *American Quarterly* for May 1976.
7. A brief glance at the principal bibliography of American publications in this period, Charles Evans' *American Bibliography* (14 vols., New York, 1904-1959) will make this change clear.
8. Julian P. Boyd, *The Declaration of Independent: The Evolution of the Text* (Princeton, 1945), 34.
9. Thomas Paine, "Common Sense," in *The Life and Major Writings of Thomas Paine* (New York, 1945), 29.

. 8 .

Intellectual History and Religious History

This paper was written for the Wingspread Conference on New Directions in American Intellectual History held at Racine, Wisconsin, in December 1977. Thus it was prepared for a small group of colleagues, most of whom knew each other, and most of whom were young. Many of the other papers, some of them brilliant, dealt with the problems of intellectual history in an age dominated by various kinds of social history, all egalitarian in their emphasis. The single point I wanted to argue was that one cannot move toward a more democratic study of American culture without giving a great deal of attention to religion, which has been at the center of most forms of popular culture in most periods.

The paper was published by The Johns Hopkins University Press in *New Directions in American Intellectual History* (John Higham and Paul K. Conkin, eds., Baltimore, 1979), pp. 105-15, and is reprinted by permission of the publishers.

What I am going to say about this subject is personal, because I think history-writing is a rather personal business. Perhaps this is especially true of intellectual history, but I suspect that it is true to some degree of all kinds of history. However many teams assemble the information, and however many computers they run it through, in the long run the decisions about what it is worth and how to present it need to be made by an individual. There are really very few successful examples to the contrary.

This is, I think, because the data do not tell us what they are for; the answers do not generate the questions. Each historian has to decide what he is trying to do, and why, and how to go about it according to everything he has learned, not just in graduate school, but in his life. Thus we will have as many visions of the past as we have historians. Gradually most historians may come to agree on some things, but whatever consensus is reached will continue to be fleeting and unstable. In every generation, at least since the Enlightenment, there have been some people who were made profoundly uncomfortable by this state of affairs, who demanded that we adopt a new method that will arrive at objective, scientific truth. But most practicing historians give up this ambition rather early and hope no more than to get a usable insight or two.

Perhaps this is tolerable because history is not a policy science; its goals are not, or not obviously, utilitarian. In some fields people have to struggle hard for consensus. Some consensus is necessary, for instance, in engineering, or in medicine. It would be nice if a little more could be achieved in economics. In some cultures, consensus has been and is imposed, and there the writing of history does not thrive. Chairman Mao found that he could not afford to let a hundred flowers bloom; we can and must. It may be that the irreducible independence of the historian explains in part what I find a poignant phenomenon: that able people still throng to graduate study in history even though they know that there are few jobs.

All this is by way of apology for offering some remarks based on my own work. My main hope is that what I have to say will prove interesting to others who have struggled with the problems of intellectual history. I can say for myself that I would rather hear another historian—even one I disagreed with—talking about what he has learned through his work than hear the same person say what he thinks everyone ought to do.

My own work, not through design but through a series of circumstantial and apparently random choices, has proved always to involve an effort to bring together religious and intellectual history, and to a lesser extent to place both against a social history background. I started with a religious history topic treated as social and intellectual history. My next major effort was an intellectual history topic that proved at all points to have a quasi-religious dimension. Most recently I have tried to treat a familiar major topic in intellectual history, the Enlightenment, arguing that it makes most sense when treated as religious history. My teaching has moved from a concentration on intellectual history to a concentration on religious history. But I have not ever been able to escape either field, and now realize that for me the two have always proved to illuminate each other.

At one point I was attracted by the suggestion that the two fields might turn out to be the same thing. That is if, with Tillich, one defines religion as ultimate concern, most intellectual history turns out to deal with religion whether it is called that or not—it deals with whatever people have found inescapably important. At the moment, however, I am not satisfied with this unification of the two fields. Both intellectual history and religious history do indeed deal with the ideals and values held to be most important by people in past time. But religious history deals with certain *kinds* of ideas and values, expressed in certain ways.

Some say that religion is concerned with the transcendent, a special kind of ultimate concern that nonreligious people do not share. But to define *transcendence* clearly seems to be difficult even for theologians, and is still harder for historians. For some people who have considered themselves religious—deists and some kinds of liberal Christians—God himself has not been clearly transcendent. For some patriots and romantic democrats, the nation or the people have had transcendent meaning. For Transcendentalists, all nature was an incarnation of spirit and thus transcended itself.

I find somewhat more helpful an insight I find in the essays of Clifford Geertz, which suggests that religion deals with symbols in a different way from other divisions of human culture. To Geertz and to some other anthropologists, all culture consists of systems of symbols. Religion involves a set of symbols endowed with ultimate authority and tremendous motivating power, whose function is to bring together a conception of the universe and a code of conduct.[1]

Sometimes this special kind of extremely powerful symbol system cannot be reduced to systematic verbal formulation. Here, perhaps, in the vast and rapidly changing field of language and symbol—a perilous jungle unfamiliar to most historians and certainly to me—one must begin one's search for a clue to the difference between religious and intellectual history. Both these deal with human culture and therefore with sets of symbols. Intellectual history in theory draws its subject matter from a great many aspects of culture. In practice, its writers tend to concentrate on those kinds of thought that have been systematically expressed in words, and in words used discursively. Religious history, on the other hand, deals with supremely powerful ideas and emotions that often cannot be expressed in this manner. Religious ideas and feelings have been expressed in many different ways: partly in words, and in words used in various ways; and partly otherwise, for instance in liturgical acts, demonstrations of devotion, or works of art.

Let us then rely for the moment on this difference in means of expression for our distinction between the two fields, bearing in mind also the somewhat tenuous matter of transcendence. Obviously this will leave a considerable overlap. Jonathan Edwards, for instance, clearly is part of the subject matter of both fields. His *Nature of True Virtue* uses words in the same way they are used by other eighteenth-century moral philosophers, seldom invoking any kinds of symbols not acceptable to nonbelieving opponents. On the other hand many of his sermons and meditations are in a completely different mode of discourse, much less approachable from outside religion. Perry Miller and his critics do not agree on what Edwards really meant by "eternal fires," and no scholar is likely to explain to us just how Edwards, walking in the fields, suddenly discovered the gentleness of God.

If we say that religion deals with many kinds of symbols and not only with words, it may be objected that precise verbal formulas like the Westminister Confession play a large part in the history of some kinds of American religion. Yet the first article of that confession declares that the light of nature is not sufficient to give the knowledge of God, and the second describes him not only as invisible but also as incomprehensible. Verbal formulas, from the Westminster Confession to the various platforms of late-nineteenth-century Fundamentalists, have usually played the role of unsuccessful efforts to dam and channel the flood of religious intuitions and emotions. The history of religion, in America as elsewhere, has to be much more than the history of creedal statements.

At the opposite end from Edwards in the religious spectrum, both Thomas Jefferson and John Dewey sometimes approach the border between religious and intellectual history. Both spent a lot of time worrying about ultimate matters and proposing new religious formulas. But neither was able to develop sufficiently powerful symbols to make his religious insight authoritative for many people. Nature's God and the *Common Faith* remain elusive in a way that the Crucifixion or the Exodus are not elusive, although the meaning of these last may in turn be difficult even for a believer to state in clear grammatical sentences. On the other hand, some people who we think of as living on the far borders of religious history have developed a pretty powerful set of symbols, which is authoritative and perhaps transcendent for them. Here one thinks especially of the proponents of Manifest Destiny, the religion of the Great Republic, whose symbols include both objects—the flag—and events—the landing of the Pilgrims. For some adherents of the quasi-religion of American nationalism, the assassination of Lincoln in the moment of victory and forgiveness has tremendous symbolic authority, clearly dependent in large part on its Christian parallel.

There is indeed a blurred border area, and yet the distinction is not entirely without its uses. For instance, the endlessly repeated emotional formulas of the revival sermon, or the ritual of the camp meeting with its mourners' bench and anxious seat belong with religious and not with intellectual history. The exercises in deductive or inductive reasoning of modern academic philosophers or sociologists belong with intellectual and not with religious history. And as these examples suggest, a major difference between the two fields, at least when one sees them in the American context, is that intellectual historians are likely to deal mainly with an educated elite, religious historians with many kinds of people on many levels of articulateness and education.

This has, I think, been a division that scholars and teachers of American intellectual history, including me, have been reluctant to admit. Part of the difficulty has arisen from the emotionally complicated relation between American intellectuals and the masses. Most academic Americans in recent times have considered themselves liberal democrats. They—or rather we—have wanted to be on the side of the people, even perhaps to love the people. But most of us have been made aware at some time in our lives, perhaps painfully, that large sections of the people do not share our tastes and opinions and do not feel that they need or want our work. Thus sometimes we have stopped looking at the people as they are and have been, and have instead constructed an ideal version of the history of American culture as it ought to have been from a liberal intellectual point of view.

I will call this version, loosely, the Parringtonian synthesis. It goes something like this: Americans of the seventeenth century were obsessed with religion, especially Calvinism. From this they were liberated in the eighteenth century by the combined influence of the Enlightenment and the Revolution. In the nineteenth century, as Americans became less European and therefore more egalitarian, they progressively turned away from supernaturalism to various kinds of religious liberalism, and eventually some of them moved beyond this to scientific naturalism. In the twentieth century, this process has gone far enough so that one need take little account of religion at all; where it survives it is mere social convention. Contemporary cultural history is expressed mainly through science, social science, and literature.

This is no doubt a caricature, and it may seem that I am setting up a straw man. Yet I would ask you to think about this mode in relation to our actual presentation to students. A good deal of it, I find, was implied in the organization and selection of topics in the course in intellectual history I taught for more than twenty years and have now happily turned over to a very able colleague. Something that always made me uncomfortable in this course, and that was sometimes noticed by bright students, was that it skipped back and forth between an attempt to deal with the community or with large groups and a concentration on elites, sometimes tiny elite groups like the Transcendentalists, or even on isolated individuals. Was a lecture on the Great Awakening or the Social Gospel really part of the same sequence as one on Thorstein Veblen or the expatriates of the 1920s? Part of the difficulty I think was that I was moving back and forth, without realizing it, between intellectual history—the history of ideas systematically expressed in words—and religious history, the history of ideas and feelings expressed in other kinds of symbols as well as in words, and in words used in various ways. More recently I have begun teaching instead a short survey of American religion, broadly defined. This too of course has its epistemological and organizational problems, and always will. But I feel a little more as though I am talking about the same thing all through the course, even though I am dealing with extremely different kinds and numbers of people. And to me, at least, the version of American culture I now present seems richer and more various.

I should like briefly to run through some of our standard topics, and to try to suggest how they look different if one moves from intellectual history to religious history, and how making this change helps to lay to rest the ghost of the Parringtonian synthesis.

In dealing with seventeenth-century intellectual history, most of us used to draw heavily on the work of one great, dramatically compelling scholar. Our lectures were likely to deal with the Puritan errand, with the covenant, and with declension. Like all great scholars, Perry Miller has attracted swarms of critics, and in his case some of them have made telling points. We have learned to question the theory of the covenant theology as a strategy shaped to deal with declension. We have learned to ask whether one has finished with the cultural history of the New England colonies when one has dealt with the ideas of the leading clergy. And we have been reminded that New England was not America. All this makes Miller's subjects, the Puritan clergy, more real and interesting. Their task becomes more significant when we realize that not everyone in sevententh-century Massachusetts Bay was obsessed full-time with religious ideas, any more than was everyone in thirteenth-century France. (Perry Miller knew this, of course, and sometimes made it clear that he knew it.) It is valuable to see the tasks of the Puritan clergy in terms of their daily round of religious duties, carried on in an environment that was precarious from the beginning.[2] The theory of decline set forth in jeremiad sermons is more interesting when it is seen not as a report on the actual situation but rather as a hortatory device. Without going back to the rather silly cynicism of some of Miller's predecessors, fully recognizing the importance of all he has taught us, we can perhaps see religion in early New England in somewhat the same way as we see religion in other periods of American history—as part of an adaptation of traditional beliefs and symbols to a new situation, a situation involving some degree of voluntarism and pluralism from the start. We may be able to argue that religion was important in the seventeenth century in New England without implying that it was unimportant in other times and places.

When we come to the eighteenth century we have the task of fitting together, in America as in the world, the Age of Enthusiasm and the Age of Reason. We have learned from many excellent scholars that the relations among evangelical Protestantism, the Enlightenment, and the Revolution are extremely complicated, and different in different American places and subcultures. We know that if many of the leading Revolutionaries were inclined towards deism, many of their most ardently committed and even violently revolutionary troops were evangelical, that many people saw the revolutionary struggle not as liberation from Calvinism but as a means of arresting moral and sometimes doctrinal decay. I have argued that parts of the Enlightenment—especially in its most antireligious aspects—were tied closely to the culture of the European upper classes and thus to those Americans who clung most firmly to that culture. Some of the Enlightenment was at once religiously radical, socially aristocratic, and politically conservative—some but by no means all. If this is so it was clearly not possible that the Enlightenment in anything like its European shape could carry with it the independent American people. As American culture became more democratic and less European, the Enlightenment could survive only in drastically adapted form. Another way of putting this might be to say that it remained an intellectual movement and failed to become a religion.

The themes of our cultural history in the early nineteenth century—perhaps the most formative period—are four: the growth of evangelical religion, social and political egalitarianism, nationalism, and reform. All these grew together and cannot be separated, and their relation with each other was complex and shifting. It has not worked, for instance, to range evangelical religion *against* either democracy or reform—it was too much the heart and soul of both. Yet it is also true that the leaders of certain important kinds of evangelical religion distrusted democracy and engaged in a no-holds-barred battle against certain kinds of reform.

To make sense of this complicated story (and I believe no one has quite done this) one has to make some use of religious distinctions and even terminology. For political and social as well as religious history one has to speak of theocrats and antinomians, millenarians and comeouters. Among the fascinating problems is the relation of that important but tiny group of self-appointed spokesmen, the Transcendentalists, to the whole national scene. It was clearly an ambivalent relation: like so many American intellectuals, Emerson and his friends were deeply on the side of the people but differed with most of them on almost every religious or political question. And yet for some odd reason Concord intellectuals were saying some of the same things as revivalist preachers and Jacksonian editors[3] about certain matters crucially important to all: the sources of knowledge and the tests of truth. Really to understand Emerson and Thoreau, or for that matter Whitman and Hawthorne, one has to bear in mind, among others, Charles Grandison Finney, Phoebe Palmer, and Joseph Smith. Of course no one can do that adequately; that is what makes our enterprise so inexhaustibly absorbing.

For the late nineteenth century almost everyone starts with the immense facts of industrialization and urbanization. Here and at many other points we must go well outside the bounds of religious or intellectual history if we are to understand either. And yet it must be borne in mind that people of this generation were reacting to these physical stimuli in terms of the complex intellectual and religious tradition carried on from earlier periods. Industrializing America is not even the same as industrial England, and utterly different from industrializing France or Germany or Russia. Part of the difference must be sought in religious history.

The country was still officially Protestant in its odd unofficial way, and leading political spokesmen saw American Protestantism as part of the coming worldwide social and political millennium. And yet this dominantly Protestant country was developing a working class a large part of which was Catholic—a situation quite special in the world. Within Protestantism a rather bland variety of theological liberalism was powerfully entrenched in the socially dominant churches.[4] From the same liberal theological principles different kinds of sincere Protestants drew political and social conclusions ranging from the most rigid laissez-faire conservatism to reformist semisocialism. This socially dominant Protestant liberalism was so powerful that it exerted a strong and disruptive pull on both Catholics and Jews. Yet in the Protestant camp itself, liberalism was increasingly challenged by orthodox revivalism and militant religious conservatism. It would be hard to say whether more people were influenced by Henry Ward Beecher or Dwight L. Moody.

More than we used to realize, in this period of social upheaval social critics made natural and fervent use of religious terminology. This was true not only of millennial socialists like Edward Bellamy but also of Populists, raised on camp meetings and revival hymns and by no means abandoning either their forms or their message.[5] It is true, and important, that on the top intellectual level various people challenged, qualified, abandoned, or fought against all the dominant religious assumptions. But one cannot understand William James, Thorstein Veblen, or John Dewey without remembering the agonies and struggles of doubters in an age when religious commitment was taken for granted.

Finally, let us consider the last article of the Parringtonian synthesis. Is religion a nearly negligible topic in American culture of the twentieth century, as many intellectuals living in that time have assumed?[6] Here I want to be careful not to over-state my case for polemical purposes. It is clear enough that religion does not occupy the same place in American culture that it did in the days of Henry Ward Beecher, let alone in those of his father, Lyman. It is indeed still part of the official rhetoric. It would probably be harder for a critic of majority religion to be elected President now than in the days of Jefferson. It is tempting to dismiss political religiosity as meaningless verbiage, but this is clearly inadequate in the days of Jimmy Carter. Even if one finds official religion meaningless, the reasons politicians find it necessary are important to think about.

On the other hand, since the First World War or a little earlier, religion has ceased to have a place in the established culture of the intellectuals. In some intellectual centers it prob-ably takes more courage for a writer or painter or professor to go to church or synagogue than not to—though neither choice involves any serious disabilities. About middle-class religion, the problem raised by Will Herberg is a real one: Is it possible in a country with several equal religions to talk about "religion in general" in a way that has any meaning? Robert Bellah has suggested that one of the powerful surviving religions is the civil religion, a suggestion that some people find depressing if true.[7] For a number of reasons then, it is plausibly and widely sug-gested that religion is now merely a meaningless camouflage for conservative patriotism, the pseudoreligion of Americanism.

I would call this a corollary of neo-Parringtonianism and suggest that it is at most only partly true. To begin to move toward a full understanding of the present complicated religious situation one would have to take into consideration a whole series of phenomena that challenge this thesis.

First, one must bear in mind the familiar statistics that show an almost steady growth in church membership, which reached its peak in 1965, and has declined only slightly since. To say that something which attracts some concrete allegiance from 60 percent of the population is insignificant one must be extraordinarily certain about what is important and what is not. Second, one must look hard at the great and rising popularity of many kinds of evangelical Christianity and pentecostalism, kinds of religion that are not anything like civil religion, and that have all sorts of different relations, antagonistic as well as friendly, to the religion of Americanism.[8] Third, one cannot omit the flourishing of the mystical cults, old and new, Eastern and indigenous, many of them centrally related to complaints against the materialism of American culture. Finally, right in the intellectual center, or at least in the intellectual centers, one has to confront the youth movement of the 1960s, which seemed to many observers to be full of echoes of the period of transcendentalism. Much of the New Left and its residues can, I believe, be seen as centrally a repudiation of all varieties of scientific naturalism and a search for transcendent values and symbols in which to express these. Naturally enough, this tendency has deeply distressed many partisans of the Old Left. To summarize: in recent times the position of religion in American culture is different and difficult, but we get nowhere in understanding it by dismissing the problem.

Now I come to the difficult question. Is it possible to draw any helpful conclusions from this sketch, which by no means pretends to offer a synthesis to rival the defunct one I have labeled Parringtonian? First, my suggestions do not dispose of the idea that the intellectual history of the United States is a story of the movement from Calvinism through various kinds of liberalism to secularism. This organization holds roughly true if one defines intellectual history as dealing with an articulate elite, and perhaps this is how it should be defined. Such a definition need not be damaging. No one would be more hostile than I to any suggestion that it is somehow immoral to study elites. I would hate to think that students in the future would cease to have to come to terms with Jonathan Edwards, Ralph Waldo Emerson, the James brothers, or Henry Adams. But those who deal with these fascinating figures and others like them will find them even more interesting if they deal with them in the full context of American culture, and a large part of this context must be provided by religious history. Were these great figures really, as some of them thought they were, spokesmen for the people as a whole? This cannot be assumed, and the argument will prove a tricky one.

For those who deal with religious history it has become apparent that a progressive synthesis, running from Calvinism to liberalism, will not do. This again works only when one deals with the most verbally articulate sections of American religion. As Sydney Ahlstrom's impressive summary demonstrates, that is only a small part of the story.

I do not intend to abandon intellectual history for religious history. Rather I am suggesting that to move back and forth between these two fields makes a stimulating and fruitful trip, provided one is conscious of the movement. To take into account both intellectual and religious history will help us understand the relation between words and other kinds of symbols, between organized ideas and those only partly organized. It will therefore help students of American culture with what I think is their hardest and most important problem, the special and unique relation between elites and masses in this country. It will open questions one does not think of otherwise. In fact it has already done this: think of the problem of civil religion and also the history of all kinds of millennialism. It may provide categories for investigating phenomena that have proved perennially elusive, including the springs of social reform in many periods.

I hope that in my enthusiasm—the enthusiasm of a convert—I am not telling anyone what he should work on, or still less what he should like. I do not think one need admire something either because it is big, or because it is as American as apple pie—or evangelical Protestantism. There is nothing wrong, for instance, with finding Ralph Waldo Emerson more admirable and interesting than Joseph Smith. But I am suggesting that it will make the study of either more interesting if one remembers constantly that the other was also powerfully present.

Notes

1. Clifford Geertz, *The Interpretation of Cultures: Selected Essays* (New York: Basic Books, 1973), especially "Religion as a Cultural System," pp. 87-125, and "Ethos, World View, and the Analysis of Sacred Symbols," pp. 126-41.
2. David D. Hall, *The Faithful Shepherd* (Chapel Hill: University of North Carolina Press, 1972); J. W. T. Youngs, *God's Messengers: Religious Leadership in Colonial New England, 1700-1750* (Baltimore, Md.: The Johns Hopkins University Press, 1976).
3. See John William Ward, *Andrew Jackson: Symbol for an Age* (New York: Oxford University Press, 1955).
4. The main ingredients in the kind of liberalism I am talking about here were divine immanence and evolutionary progress, sometimes associated with a social interpretation of the New Testament promises.
5. For a valuable new survey of Christian semisocialism, see Peter J. Frederick, *Knights of the Golden Rule* (Lexington: The University Press of Kentucky, 1976). A useful beginning is made in the neglected subject of Populism and religion by Robert C. McMath, *Populist Vanguard* (Chapel Hill: University of North Carolina Press, 1975), esp. pp. 62-76, 133-37.
6. A representative early example is Harold E. Stearns, whose *Civilization in the United States* (New York: Harcourt, Brace, 1922) looks at that subject from the point of view of the young intellectuals. In his preface Stearns tells us that when he tried to procure an article on religion his prospective contributors told him "almost unanimously" "that real religious feeling in America had disappeared, that the church had become a purely social and political institution, that the country was in the grip of what Anatole France had aptly called Protestant clericalism, and that, finally, they weren't interested in the topic" (p. vi).
7. Bellah's much discussed essay on civil religion and a number of the essay's critics are presented in Russell E. Richey and Donald G. Jones, eds., *American Civil Religion* (New York: Harper & Row, 1974).

8. A helpful taxonomy of recent evangelical religion is offered by Richard Quebedeaux in *The Young Evangelicals* (New York: Harper & Row, 1974). This is brought up to date by the same author's "The Evangelicals: New Trends and New Tensions," *Christianity and Crisis* 36, no. 14 (20 September, 1976): 197-202.

· 9 ·

The Religion of the Republic

This was given as the Faculty Research Lecture in Berkeley in 1981. These lectures attract a highly intelligent but unspecialized audience. I also intended this lecture to serve as the opening chapter of a book of essays on religion in American society. Other chapters of this book would deal with the important kinds of American religion that remained outside the dominant variety described here. One essay would treat the continuing and important strain of religious pessimism, and others the recurrent outbreaks of personal, non-social forms of religion, often antinomian and mystical. Thus my attempt to define and describe the dominant national religion would be placed against a background of major dissident varieties.

In recent years some historians of American religion, reacting against earlier exclusions and prejudices, have maintained that it is meaningless to talk about any dominant or "mainstream" religion in the presence of so much variety. I remain convinced that through the nineteenth century there was indeed a dominant kind of religion, though it cannot be defined in denominational terms. People who were wholly or partly outside the dominant camp, like Catholics and Jews and premillennialists, were always conscious of their outsider positions. People within the dominant kind of religion often discussed dangers to it, but seldom questioned its existence.

Since giving this lecture I have made one correction of fact and several of emphasis, and have slightly expanded the footnotes. I am indebted to Professor William Hutchison of the Harvard Divinity School for helpful criticism. The lecture has not been previously published.

In this lecture I want to ask the question, "What has been the national religion of America?" I do not mean here only the religion of the churches, nor only what Robert Bellah has brought to life before us as the civil religion of the country. Rather I want to look for something that includes and transcends both of these.

According to the anthropologists of religion, a tribal or national religion has certain crucial functions, both individual and social. For a society, a national religion must confirm its values, define its loyalties, legitimate its institutions, resolve its conflicts, sanctify its triumphs, explain and thus render bearable its disasters. At the same time, for the individuals who make up the society, a satisfactory religion must provide meaning for their lives, order and dignity for their rites of passage, and some sort of solace or at least meaning in the face of death. I am much impressed by Peter Berger's suggestion that perhaps the final function of religion comes down to theodicy—technically, the vindication of God's justice; more broadly, explaining why the universe in which we live frustrates so many of our hopes. Happiness, says Berger, is not essential for human beings, but meaning is essential, and in most societies, meaning has been the province of religion.[1]

Obviously and logically, these immense tasks of a national religion can best be performed in a situation of religious establishment, or, still better, of taken-for-granted religious unity. In many societies in many periods, articulate religious dissent has been not so much prohibited as unthinkable. Clearly, pluralism and secularism would seem to diminish the status of religion, rob it of its authority, and get in the way of its functions. America has certainly been the country of pluralism and some would say of secularism as well. Yet such acute observers of America in the nineteenth century as Alexis de Tocqueville and Philip Schaff reported to astonished Europeans that religion in America, fragmented and unsupported by government, was healthier than religion in Europe. I think that this was true during the whole of the nineteenth century, the period when America formed its characteristic way of life and rose to world power. While religion was under powerful attack in several major industrial countries, it continued to pervade and sustain American culture. People in America continued to take it for granted that religious and national values were closely related and indeed almost indistinguishable from each other.

What was this powerful national religion of nineteenth-century America? That is the question this lecture seeks to answer. Not what American national religion ought to have been, or what we wish it was, but what it was in fact.[2] First it is necessary to say briefly what it was *not*.

One early claimant for the status of national religion was deism, the religion of enlightened thinkers of the eighteenth century, including some—by no means all—of America's founding fathers. Deists believed that the world had outgrown the puzzling fables and contradictory beliefs of the Bible, and that intelligent people should worship a single beneficent God, who had created a comprehensible world and was running it through the laws of nature. Thomas Paine, for one, was sure that the new republic, having overthrown irrational political institutions, would finish the job and get rid of its irrational religious institutions as well.

Thomas Jefferson believed that the beneficent light of reason would eventually prevail over the last-ditch defenders of Trinitarian Christianity, though he thought that the New England clergy, whom he hated, would be the last to give in. From Jefferson's day to that of John Dewey and beyond, liberal thinkers have periodically proclaimed, in vain, the fast approaching triumph of the religion of reason, or science, or humanity. For a moment, nearly at the end of the eighteenth century, some of the defenders of orthodoxy agreed with their enemies, and feared that deism was close to triumph.

All these hopes and fears proved equally illusory. By very early in the new century one kind or another of Biblical religion was everywhere triumphant and it was the deists who were reduced to little sectarian groups, bravely resisting extinction. Had that eighteenth-century sceptic, Edward Gibbon, lived long enough, he would surely have found in the nineteenth-century United States his most telling example of the triumph of barbarism and Christianity. Deism and its liberal successors, always proclaimed as the religion of the republican future, never became the religion of more than a small elite group.

One reason for the failure of deism arises from one of the principal facts of American social history, one almost never understood by Europeans, in the eighteenth century or later. In Europe, the principal churches were identified with the aristocratic and monarchical order, increasingly seen as oppressive. In America, the churches were popular institutions, identified ever since the exodus from Europe with resistance to oppression. Moreover, in America as elsewhere, deism failed to develop either of two essentials for a successful popular or national religion. Despite a few sporadic efforts, deists never put together an adequate ritual. Even more important, deism was unable to develop a theodicy, and had no explanation for the existence of evil. If both God and man were essentially benevolent, why was the world as it was? Enlightened eighteenth-century gentlemen might say that the modern world was rapidly becoming more predictable, cheerful, and comfortable. This was not a compelling argument for those who constantly found their crops destroyed by freaks of weather, or their children carried off by the epidemics prevalent in the towns and on the frontier. For people subject to privation and periodic catastrophe, even a punishing God is more comprehensible than an entirely beneficent one. Deism could carry out some of the functions of a civic religion adequately, but it could not succeed as a personal religion. One finds in the letters of some eighteenth-century Virginians brave efforts to face the death of their children or spouses as deists, but these are not very successful. The cool preachments of the religion of reason offered little at times either of sorrow or joy.

Yet deism played some part in the development of America's national religion. In odd alliance with sectarian pietism, it helped in the period after the Revolution to complete the achievement of religious freedom, the special pride of the republic. Moreover, since it played a major part in the civic religion preached by some of the founding fathers, it could not be altogether extirpated from civic observances. On every Fourth of July, the Declaration of Independence, a covertly deist document, was read after an explanatory invocation by a Presbyterian or Baptist clergyman. Some Americans, for a while, were stuck with the difficult combination of a deist civic religion and a Calvinist personal religion. And deism, at the end of the eighteenth century, played the essential role of a major antagonist for orthodox Christianity. Religions, like ideologies, are deeply affected by the enemies they choose, and the dominant American religion, having denounced and defeated the deists, had to insist that it too was entirely compatible with progress and with patriotism.

Until about 1800, this dominant American religion was Calvinism. It had never been exactly the Calvinism of Calvin, but rather that of Dort and Westminster, emphasizing the absolute omnipotence and inscrutability of God, the total depravity of man, and the logical necessity of predestination to heaven or hell. Out of this apparently unpromising material, seventeenth-century

New England constructed the most effective tribal religion that ever existed in white America. New England Puritanism was highly organized, though its loose congregational structure fitted the needs of an expanding society. Its explanation of the justice of God was logically satisfying though by no means cheerful. Its determinism never led to quietism, but like Marxist determinism seems somehow to have spurred many to constant effort.[3] Its austere rituals seem to have helped many to accept the facts of life and death. For the community it developed the supreme ceremony of the jeremiad, in which a selected preacher pointed out to the people, in the solemn presence of their elected leaders, the great destiny and sorry failings of New England.

As its many fine historians have shown, New England Calvinism was constantly adapted and altered throughout its long life. From the beginning preachers emphasized the more reassuring parts of its message without quite denying its severe premises. Especially in the two great revivals of the 1740s and of 1800 its mood and tone were changed by popular emotional presentation. Calvinist ministers gained prestige by their almost unanimous support of the American Revolution, and then lost some of it by their narrow and bitter attack on the French Revolution and Jefferson. In the expansive years of the early nineteenth century, Calvinist doctrines were adapted by theologians of great skill. The social teachings of Calvinism were reconciled with voluntarism, democracy, and reform by that gifted and dramatic publicist Lyman Beecher, alternately groaning and exulting every inch of the way.

Despite all efforts to sustain and adapt it, by about 1815 Calvinism in any recognizable form was no longer the dominant American religion. Yet it is wrong to underrate its lingering hold on American nineteenth-century culture. As one of its excellent historians says:

> Those who see only the optimism and innocence of Jacksonian America miss the tough realism, the sense of human fallibility, that informed the post-Calvinist religious temper.[4]

The purest Calvinist doctrines were defended with great skill and vigor at Princeton and elsewhere. In its adapted form, Calvinism was able to fight to a standstill the Unitarianism of Boston, the newest form of Enlightened liberalism. In somewhat degenerate form, Calvinist doctrine helped to form the intellect of most of the American elite, since it was taught at most of the colleges. Here it was combined with laissez-faire economics to show the necessity of poverty and the depravity of the lower classes. Still more important, and far beyond the boundaries of any formal doctrinal allegiance, Calvinist ideas of child-raising continued to form the personality of countless Americans, taught from infancy to fear and control their impulses.[5]

On the highest intellectual level, Calvinism gave much of what depth it had to American nineteenth-century high culture.[6] Sometimes Calvinists held out against racism, insisting that all men were equally depraved, and pointed to unpleasant and difficult social duties, such as abolishing slavery.[7] Even in mid-century, Calvinist doctrine often seemed to be reconcilable more honestly than its liberal competitors with the teachings of Darwinism. In the greatest age of New England literature, some of the most profound thinking and writing was influenced, directly as well as indirectly, by what Herman Melville, writing about Nathaniel Hawthorne, called

> . . . that Calvinist sense of Innate Depravity and Original Sin, from whose visitations, in some shape or other, no deeply thinking mind is always and wholly free.[8]

For all its important residues, however, Calvinism was both too austere and too difficult to be the national religion of America in the age of dramatic expansion and romantic emotion. By the beginning of the nineteenth century, Calvinism was everywhere declining, evangelicalism or revivalism rising. Evangelical religion is the emotional and fervent kind that depends on spreading the gospel, on individual conversion and commitment rather than on church control or theological argument. Revival-

ism is the social counterpart, in America, of this deeply individual kind of religion: the technique and practice of arousing collective religious enthusiasm.

In the early nineteenth century, revivalism spread through the
whole culture of America, going far beyond religion into a host
of causes good and bad, from temperance and antislavery to anti-
Catholicism. The hope of immediate salvation and transformation spread from the individual to the national soul. A writer
of 1829 quoted by Perry Miller explains this:

> The same heavenly influence which, in revivals of religion,
> descends on families and villages . . . may in like manner,
> when it shall please him who hath the residue of the Spirit,
> descend to refresh and beautify a whole land.

"The whole land"—Perry Miller repeats, overwhelmed by the
idea—"the whole beautiful, terrible, awesome land!"[9] As another
excellent writer puts it, revivalism became America's answer to

> a whole host of problems: barbarism on the frontier, infidelity
> among the intelligentsia, panic in economic crises, obduracy of
> Southern slaveholders, and, most importantly, expansion of
> Protestant churches throughout the growing nation.[10]

As it spread, revivalism developed its own loose but effective
ritual, with such devices as the mourners' bench and the anxious
seat, as well as its increasingly practical methods, the advance
team, the protracted meeting, etc. Its theodicy was simple and
satisfactory: man was everywhere sinful, but nowhere without
hope, and God's plan included the evangelization of the world in
our time. In its different Eastern, Western, and Southern forms,
it offered the kind of social control badly needed by the turbulent and expanding nation. Not by establishment, but by subtle
methods of inquiry, discipline, and ostracism it could rebuke
drinkers, wife-beaters, cheaters, brawlers, and roughnecks.[11]

A major element in American revivalist religion was millennialism, the ancient Christian expectation of the immediate ap-

proach of the end of the world, followed by the thousand-year reign of peace and love. Closely related was perfectionism, an idea first arising out of Methodism and spreading beyond it. According to perfectionists the individual soul, after conversion, can and must grow in grace until it is completely free from sin. And as each individual soul moved toward perfection, so must the nation and the world. By the middle of the century many American Christians believed in and worked for the imminent achievement of a world without drink, war, sin, or slavery.[12]

Thus by 1850 or earlier all the elements were in place that went to make up the religion of the major and growing American Protestant churches. Still firmed up by the remnants of Calvinist discipline, American Protestantism of this period tended to be voluntaristic, activist, moralist, revivalist, missionary, and emotional. It was individualist yet conformist, egalitarian but not revolutionary, millennial and perfectionist, and utterly un-European.[13] American revivalist religion was also often narrowly moralistic and anti-intellectual, and its pervasive method tended to turn every cause, good or bad, into a rapidly growing and soon fading popular crusade. Nonetheless, it seemed to fit perfectly the feelings of most of the people of the great republic in the age of easy expansion and romantic emotion.

This was the dominant religion of the American *churches* in the middle of the nineteenth century. But our subject today is not just the religion of the churches but the religion of the people and of the nation. This I think was put together through a combination of evangelical Protestantism with American nationalism, with its Enlightenment roots and its romantic flowering. Let us call this national religion Progressive Patriotic Protestantism, and let me try now to explain what I mean by these three words.

The national religion was *progressive*. The belief in secular progress inherited from the Enlightenment was completely assimilated to the evangelical preaching of the dawning millennium. The millennium itself was to be achieved by the rapid conversion of the world to democracy and Protestant Christianity.

The national religion was *patriotic*. America, with England sometimes admitted to partial partnership, was the pattern for the world, the destined theater of the millennium, the chief motor for universal conversion.

The national religion was *Protestant*. Its spokesmen constantly invoked the mythology of the Protestant past. The forward march of freedom had commenced with the Reformation and flowered with and after the American Revolution, providentially distinguished from the godless French Revolution. Salvation by faith and the priesthood of all believers were doctrines valid for both the secular and the sacred spheres, and were sometimes translated into romantic adoration of the popular instinct. Personal and sexual morality, long the special province of the evangelical churches, were crucially important for the success of the national mission.

One must quickly admit that Progressive Patriotic Protestantism was never the religion of the whole people. It was, however, the vision of the crucial and dominant Northern middle class, a group which often forgot that it was *not* the whole people. It was the religion of those who dominated the biggest and richest churches, the national religious press, the interlocking reform movements, the colleges, the national magazines, and to some extent the politics of the nation.

The hegemony—to borrow a helpful Marxist term—of Progressive Patriotic Protestantism began about 1815 and ended exactly in 1919. Of course this national religion developed and changed during its long triumph. The mission of America to be a pattern to the world was clearly foreshadowed in some of the propaganda of the Revolution. Faltering during the dangerous divisions of the 1790s, that mission became manifest about 1815, with the end of the war with England, the defeat of deism, the triumph of the Great Revival, the beginning of worldwide Protestant missions, and the revival of the westward expansion of population. To quote one of the best historians of early America, by

1820 "the republican citizen had become a Christian demo-
crat."[14] From then on, American isolation, territorial expansion,
and economic growth came to seem part of the Divine Plan,
which as it unfolded became nearly identical with the national
mission.

Until about 1850 this national religion, full of the ardor of
the evangelical present, yet retained some flavor of the Calvinist
past: the specifically Protestant ingredient held its own against
the other two. In the midst of millennial rejoicing, many prophets and preachers, especially Easterners with Calvinist roots,
sounded the note of the Puritan jeremiad: great opportunities
bring with them great perils and great responsibilities. The
principal danger to the worldwide triumph of American principles was that of moral decline in America itself. The destined
nation was threatened by the growth of violence, selfishness, and
impurity. Fortunately, however, America was protected by a
powerful set of institutions established by providence: universal
suffrage, the free school, the open Bible, the Christian Sabbath,
the Christian mission, the revival, the reform organization. Active evangelism and above all hard work were the motors of
destiny. Among Westerners and non-Calvinists, while the absolutely central place of morality was never denied, the millennial
future itself received more emphasis than the dangers to it. A
few quotations will help to convey the flavor of the national religion in this period in some of its shadings:

From a Sunday-school life of George Washington, 1827:

> Young Americans! as you grow up to manhood and enjoy the
> great blessing of freedom from all unjust and oppressive laws
> of man, beware of wishing to be free from the just and righ-
> teous laws of your Creator.[15]

From a biography of an evangelist, written by a conservative
Calvinist in 1841:

> In this country, the State cannot use the public treasure in
> advancing Christianity, but that every statesman and ruler,

and judge, should be a Christian in all his conduct, private and official, and should particularly be a zealous and liberal patron of Home missions, is demanded alike by patriotism and by religion.[16]

From Lyman Beecher, the principal prophet of modified Calvinism:

The time has come, when the experiment is to be made, whether the world is to be emancipated and rendered happy, or whether the whole creation shall groan and travail together in pain. . . . If it had been the design of Heaven to establish a powerful nation, in the full enjoyment of civil and religious liberty, where all the energies of man might find full scope and excitement, on purpose to show the world by experiment, of what man is capable . . . where could such an experiment have been made but in this country?[17]

And finally, from Andrew Johnson, later President, in his ardent Jacksonian youth:

I believe man can be elevated; man can become more and more endowed with divinity; and as he does so he becomes more God-like in his character and capable of governing himself. Let us go on elevating our people, perfecting our institutions, until democracy shall reach such a point of perfection that we can acclaim with truth that the voice of people is the voice of God . . . . [the United States] has undertaken the *political redemption of man,* and sooner or later the great work will be accomplished. In the political world it corresponds to that of Christianity in the moral. They are going along, not in divergents nor in parallels, but in converging lines— the one purifying and elevating man religiously, the other politically . . . .[18]

In the revival of 1857 doubts and fears appeared forgotten as moralism, manifest destiny, and business methods blended in apparently unopposed and triumphant harmony. And finally, with the coming of the Civil War, after an initial period of hesitation most of the Northern clergy and the moralistic, reformist portion of the laity rallied to the cause in a burst of Old Testament patriotism. The war against slavery was the last bloody test of the chosen Christian nation, its outcome a final and almost complete sanctification.[19]

In the late nineteenth century, a supremely optimistic and often sentimental religion triumphed over the last remnants of Calvinism in the churches of the urban middle class. Sometimes it coalesced perfectly with the cheerful individualism of triumphant business, sometimes it preached the necessity of reforming capitalism in the direction of Christian brotherhood. Together, most social conservatives and most social reformists perceived their century as a triumphant march forward of American society and American religion.

As earlier, warnings of possible dangers were sometimes sounded, especially in the recurrent periods of depression and social crisis. The manifest mission of the great republic might be placed in danger by cities, immorality, excessive greed, new forms of atheism, the rise of Catholicism or socialism, and especially by immigration. Yet these warnings, often sounded, were drowned out by the rising chorus of exultation over the growing size and wealth of the nation and of its churches. One major summary of the state of American Christianity published in 1888 concluded that religion might not dominate American life as perceptibly as it once had. This was

> because Christianity has largely transformed Christendom morally, intellectually and socially: and, therefore, it does not look so bright on the new background as on the old. . . . Piety has come out of the cloisters and gone forth among the masses, in imitation of "Him who went about doing good." No previous age can parallel in magnitude, in grandeur, in intelligent apprehension, the religious activities of this age.[20]

Henry Ward Beecher, the most popular spokesman of middleclass Protestantism, spelled out this rather startling comparison of the present to even the most sacred ages of the past:

> "Are we wiser than the Apostles were?" I hope so. I should be ashamed if we were not.[21]

There were of course many alarmed nativists and extreme racists among the clergy and laity of the country. But among the leading Northern spokesmen of the national religion the racism was likely to be relatively benign and the nativism patronizing rather than virulent. By the slow, gradual operations of free institutions and free religion, blacks, Indians, and immigrants would gradually be made over in the image of progressive Anglo-Saxon Protestantism. It did not always matter even that the country was no longer, strictly speaking, a Protestant country at all. The Jews, never as frightening as Catholics to America, could be seen as grateful refugees from oppression, struggling upward in imitation of earlier pilgrims. Even the Catholics were sometimes welcomed into halfway membership in the national cult. According to another major summary of American religious progress, this one dated 1893:

> They have become as American—at least the body of them—as the Lutherans[!] No impartial and intelligent person now believes that they want to subvert our liberties or destroy our government. We may justly accuse them of meddling too much at times in party politics; we may deprecate the favor they sometimes receive in municipal councils; but in all those fundamentals which make our government thoroughly and securely Republican, Catholics are at one with Protestants.[22]

Some Catholics and some Jews accepted this associate membership with great enthusiasm. At the end of the century Leo XIII felt it necessary to crack down on the most extreme Americanizers in the Catholic hierarchy. It was fine, the Pope conceded, for American Catholics to be patriotic Americans. But they could not, while remaining Catholics, say that the whole world must and would eventually imitate American religious and political institutions.[23] Among Reform Jews some spokesmen caused deep anguish to their more nearly orthodox coreligionists by asserting flatly that now America is our Zion and Washington our Moses.

For the churches and for the nation, the summit of idealism and optimism was reached in the first decade and a half of the twentieth century. Foreign missions were flourishing as never before, with Americans contributing the overwhelming bulk of the men and the money. With open frontiers, new means of transportation and communication, the protection of friendly Christian empires, the money and power of American churches and philanthropists, Providence seemed to be opening new ways for a speedy transformation of the world.[24] And at home, many veteran preachers of the Social Gospel believed that progressive reformers were achieving a new and essentially Christian synthesis of individualism and brotherhood. Washington Gladden, the grand old man of liberal Christian reform ended his autobiography, published in 1909, with an entirely typical review of the progress in reform, philanthropy, science, democracy, and world peace that he had seen in his long life and a salute to the present and future. There were doubtless struggles ahead but they would be won:

> With all those who believe in justice and the square deal, in kindness and good will, in a free field and a fair chance for every man, the stars in their courses are fighting, and their victory is sure.[25]

The most effective preachers of the national faith in this period, however, were not ministers but statesmen. Almost every major leader sounded its note. Theodore Roosevelt, a man of no great personal piety, voiced an enormously popular version of strenuous Christianity:

> If we read the Bible aright, we read a book which teaches us to go forth and do the work of the Lord. That work can be done only by the man who is neither a weakling nor a coward, by the man who in the fullest sense of the word is a Christian.[26]

And in 1912 Roosevelt himself was defeated by the greatest spokesman of the national faith, a man who embodied all three of its elements and reflected all of its history.

Woodrow Wilson, a devout Presbyterian and the son of a minister, had deep ancestral Calvinist roots. Midway in his career he had come to accept the progressive vision. He never doubted the destiny of America to lead the world by her example. Above all, he was an absolute master of evangelical rhetoric. His mastery was not that of a charlatan, but that of a deeply dedicated believer, sure that the nation was rising out of the slough of materialism into the pure air of Christian idealism.

Unlike many national or tribal religions, that of America had ordinarily been pacific. War, to most spokesmen of Progressive Patriotic Protestantism, seemed obsolete, unnecessary, bad for business, and rooted in the evil past of kings and aristocracies. It had been one of the most frequent boasts of pious patriots that America spent more for education than for armament. Only the two great wars for the survival of the nation, the Revolution and the Civil War, had been fully sanctified by American religion. In the lesser wars of 1812, 1846, and 1898 the churches, like the laity, had been divided.[27]

In 1914 when world war, almost unbelievably, challenged the hopes of the progressive generation, most believers in the national religion were convinced, like Wilson himself, that it was America's duty to help without becoming involved, to stand as an island of peace and progress, and to help bring about a negotiated peace. "May we not look forward," Wilson asked in 1915, "to the time when we shall be called blessed among the nations, because we succored the nations of the world in their time of distress and dismay?"[28]

Two years later when, as Wilson and probably most of the nation saw it, America's efforts for peace had failed, the President asked for a declaration of war in a speech that must rank as the most powerful sermon of the national faith ever preached:

> . . . the right is more precious than peace, and we shall fight for the things which we have always carried nearest our hearts, —for democracy, for the right of those who submit to authority to have a voice in their own governments, for the rights and liberties of small nations, for a universal dominion of right by such a concert of free peoples as shall bring peace and safety

to all nations and make the world itself at last free. To such a task we can dedicate our lives and our fortunes, everything that we are and everything that we have, with the pride of those who know that the day has come when America is privileged to spend her blood and her might for the principles that gave her birth and happiness and the peace which she has treasured. God helping her, she can do no other.[29]

It is not surprising, when one thinks of the historic overtones of this speech, that hard-boiled politicians wept. It is not surprising either, that the Wilsonian crusade turned out to herald the end, and not the final glory, of the national religion.[30]

This was, of course, by no means apparent right away. It is notorious that Wilson's call for a war without vengeance and recrimination brought on instead an orgy of hate and xenophobia, in which some religious leaders took leading parts. It is easy enough to find the sermons insisting that Jesus himself, if he were here, would be willing to plunge a bayonet into the guts of a German. But sober historians have recently made it clear that hysteria was by no means universal, that most of the major leaders of the American churches made some effort at restraint. What was far more nearly universal was the mood of millennial hope, the belief that the war really would finally achieve America's historic mission: to end war by establishing universal democracy on the American pattern.[31]

This mood continued into the immediate postwar period. Prohibition seemed a major triumph of an old cause backed by religious reformers. The League of Nations seemed to incarnate the triumph of democracy and thus of peace. And in 1919 the American churches launched, for the last time, an organized effort for the immediate evangelization of the whole world.

According to its excellent historian, the Interchurch World Movement of 1919 "gave climactic expression to the crusading spirit, the lofty idealism, the optimism and popular organization which had long characterized popular Protestantism." According to its principal spokesman, the movement was "the greatest program undertaken by Christians since the days of Apostles." It was supported by 60 percent of American church members, by such major lay figures as President Wilson and General Pershing, and by the country's press. Its immediate objective was to raise three hundred and thirty million dollars by 1920. Its methods were a synthesis of big business, wartime propaganda, missions, and revivals. By the spring of 1920 it was an obvious and abysmal failure. Still in a mood of hyperbole, some of its members saw this fact as "the greatest tragedy in the history of the Christian church."[32] Whether or not we accept this estimate, I think the demise of the Interchurch World Movement marked the end of the dominance in America of Progressive Patriotic Protestantism.

The immediately succeeding period, the twenties and thirties, is seen by one leading church historian as the low point of American religion.[33] Within the dominant Protestant churches, funds and enthusiasms dried up, and the main news became the bitter division between strident fundamentalists and outraged liberals, while the Menckens and the Sinclair Lewises watched in delighted fascination. Jews and Catholics, angry at partial exclusion, refused any longer to accept associate membership in American national religion. In the nation, militant idealism temporarily seemed to disappear, with the failure of the League, the overwhelming repudiation of Wilson, and the widespread evasion of prohibition. A little later the great depression of the thirties was the first American slump that saw no major revival, and the New Deal was the most secular movement of reform in American history. For better or worse, no war since World War I has been fully sanctified by American religious leaders. In World War II many were still in sackcloth and ashes for their utterances in 1917, by the Korean War few were deeply stirred, to the war in Vietnam very many were in opposition.[34]

I do not mean to say that religion in America went into a permanent state of decline, or even that the national religion I have been describing disappeared entirely. Pronouncements made in the thirties about the secularization of American life proved as premature as similar pronouncements made in the 1790s. Whether the last few decades have seen a decline in American religion is an infinitely complex and hotly debated question, with the answers largely dependent on who counts what.[35] And sometimes—for instance during such archaic rituals as national political conventions, one hears speakers intoning the exact traditional formulae of Progressive Patriotic Protestantism.

But the preachers lack the old confidence, and the audiences lack the old unity. To say just why is not easy; let us sum it up by saying that the three components have come apart. Many Protestants are not progressive; many others are not patriotic. Many patriots are not progressive. Many progressives are not even in the broadest and most inclusive sense Protestant. The deep consensus embodied in the old national religion is gone.

How is one to look at this important religion of the past from where we sit? If I have seemed at any point nostalgic for the days of Progressive Patriotic Protestantism, I have been unable to express my true feelings. Historical nostalgia is never any use. But neither is historical polemic. Denunciations or caricatures of past ideas are useless and vulgar, and in this case far too easy.

From the point of view of traditional Christianity, any kind of national religion is false, and Progressive Patriotic Protestantism pretty close to the ultimate heresy. Some American Christians always said this. Many people in the churches, both liberal and conservative, believe for different reasons that the decline of this particular national faith has been a great gain for American religion. From the point of view of liberalism, the national religion was full of racial, religious, and national arrogance. From the point of view of conservative religion, it was far too optimistic about the future of the world. And one must remember that it was never the religion of everyone, but only of the dominant middle class in its confident days. At the top of the social and educational scale, religion was sometimes more critical; lower down, what prevailed was fervent individual piety, either ecstatic or very gloomy. The pentecostal and fundamentalist movements that have flourished in recent times can be seen as part of a lower-class protest against liberal middle-class hegemony in the churches.

On the other hand, for a long time Progressive Patriotic Protestantism managed to combine with some success the functions of both a civic and a personal religion. Its Protestant content included a code of moral behavior that millions tried sincerely to follow. As for its civic component, at certain points the conviction that Americans were good made both individuals and the nation behave better than they otherwise might have. One way to judge this national religion might be to compare it with the national religions of other nations in the nineteenth century—of Britain, or France, or Germany for instance. One might guess that its democratic base made the American national religion more benign in some respects, but also more successful in disguising some kinds of oppression.

The end of a national religion, whatever its limitations, is nothing to be taken lightly.[36] Though this one has been declining sharply ever since 1919 it is still not clear what will replace it. Perhaps, unlike most societies, America will learn to get along without any national religion at all. Or perhaps a new national religion is developing. There are several powerful claimants, the most obvious of which is the triumphant and growing movement of the religious and political right. Clearly, this carries some historical echoes of the nineteenth-century national religion I have been discussing, but it is also very different. Its theological roots are full of gloom and doom, and its social message is defensive and reactionary. According to many of its principal spokesmen, America, instead of leading the world in an inevitable march toward a better future, must hold the line against powerful and evil forces. Instead of spreading the message of progress it must reassert ancient values which have been widely forgotten.

Will this new religion of the new right become the national religion? I hope not and despite recent events I do not think so. In the first place, one must remember that in this immensely diverse country, this kind of religion does not prevail outside the camp of theologically conservative Protestants, and that it cannot by any means claim the allegiance of all those within that camp. An essentially pessimistic view of history is still hard to sell to much of the American public. And in my opinion a national religion does not grow out of the strident assertions of a self-conscious movement. It must rest, rather, on a consensus so deep in the national culture that it hardly needs expression. This was the case with Progressive Patriotic Protestantism, the dominant American religion of the nineteenth century.

But prediction, gloomy or cheerful, is not the business of the historian. Fortunately, the future lies beyond the scope of this lecture, and certainly beyond the scope of this lecturer.

Notes

1. Peter L. Berger, *The Sacred Canopy* (New York, 1967; paperback ed., 1969), 58. The various works of this sociologist have been especially helpful to me.
2. This question cannot be discussed without reference to Robert Bellah's 1967 article on civil religion and the long discussion it began. The Bellah article and some valuable criticisms of it can be found in Russell E. Richey and Donald G. Jones, *American Civil Religion* (New York, 1974). Over many years my ideas on this and related subjects have been irrevocably influenced by the various works of Sidney Mead. An article and a book by Robert T. Handy explore this problem from a viewpoint similar to mine: "The Protestant Quest for a Christian America, 1830-1930," *Church History*, XXII (1953), 8-19; *A Christian America* (New York, 1971). My own suggestions in this essay can perhaps be seen as an effort to bring together the insights of Bellah, who discusses a purely civil American religion, and Handy, who discusses Protestantism as a national religion.
3. Cf. Reinhold Niebuhr, *Reflections on the End of an Era* (New York, 1934), 130-31.
4. Richard D. Birdsall, "The Second Great Awakening and the New England Social Order," *Church History*, XXXIX (1970), 364.
5. Cf. Philip Greven, *The Protestant Temperament* (New York, 1977), 22-148.

6. This will be argued in a forthcoming work by Bruce Kuklick.

7. Michael C. Coleman, "Not Race, but Grace: Presbyterian Missionaries and American Indians, 1837-1893," *The Journal of American History*, LXVII (1980), 41-60. Early Calvinist antislavery is dealt with in Oliver William Elsbree, *The Rise of the Missionary Spirit in America, 1790-1815* (Williamsport, Pa., 1928).

8. Melville, "Hawthorne and his Mosses," reprinted in Edmund Wilson, *The Shock of Recognition* (New York, 1943), 192.

9. Miller, *The Life of the Mind in America* (New York, 1965), 11. A little later in the same book (p. 57), Miller has another striking insight about the whole missionary and revivalist thrust of the period: "We come to an inner, if not *the* central, mainspring of the missionary exertion as we recognize it as a form of romantic patriotism."

10. C. C. Goen, "The 'Methodist Age' in American Church History," *Religion in Life*, XXXIV (1965), 565.

11. Examples of church discipline in the West can be found in T. Scott Miyakawa, *Protestants and Pioneers* (Chicago, 1964). For Southern church discipline, see David Bailey, "Slavery and the Churches: The Old Southwest" (unpublished dissertation, Berkeley, 1979).

12. See Timothy L. Smith, *Revivalism and Social Reform* (New York and Nashville, 1957).

13. Un-European, but not entirely un-English. The best analysis of the similarities and differences between English and American revivalism is Richard Carwardine, *Transatlantic Revivalism* (Westport, Conn., and London, 1978).

14. Gordon S. Wood, ed., *The Rising Glory of America: 1760-1820* (New York, 1971), Introduction, 14.

15. Anna Reed, *Life of George Washington* (Philadelphia, American Sunday School Union, 1827), 209; quoted in Jacqueline Reinier, "Attitudes toward and Practices of Childrearing, Philadelphia, 1790-1830" (unpublished dissertation, Berkeley, 1977).

16. "Rev. Dr. Skinner" (otherwise unidentified), introductory essay in George B. Cheever, *God's Hand in America* (New York, 1841), xiv-xv. This book is an excellent example of the conservative variety of the national religion in the early nineteenth century.

17. Lyman Beecher, "The Memory of our Fathers" (1827), quoted in Sidney Mead, "American History as a Tragic Drama," *Journal of Religion*, LII (1972), 336-60.

18. Quoted in Winthrop S. Hudson, "The Methodist Age in America," *Church History*, XLIII (1974), 8-9.

19. This is convincingly demonstrated in James H. Moorhead, *American*

Apocalypse: Yankee Protestants and the Civil War, 1860-1889 (New Haven and London, 1978).

20. Daniel Dorchester, *Christianity in the United States* (New York, 1888), 698-99.

21. Beecher, *Life of Jesus the Christ* (New York, 1871), quoted in Paul A. Carter, *The Spiritual Crisis of the Gilded Age* (DeKalb, Illinois, 1971), 166.

22. Henry King Carroll, *The Religious Forces of the United States, etc.* (vol. 1 of the *American Church History Series*, New York, 1893), Introduction, lix-lx.

23. The standard accounts of this fascinating episode are Thomas T. McAvoy, *The Great Crisis in American Catholic History* (Chicago, 1957); and Robert D. Cross, *The Emergence of Liberal Catholicism in America* (Cambridge, Mass., 1958).

24. See, for instance, the summary of present opportunities for the spread of the gospel at the end of William Speer, *The Great Revival of 1800* (Philadelphia, 1872; rev. ed., 1903), 92-93.

25. Gladden, *Recollections* (Boston, 1909).

26. Roosevelt, Speech to the Long Island Bible Society, Oyster Bay, quoted in Charles E. Kistler, *This Nation under God* (Boston, 1924), 185.

27. On the War of 1812, see William Gribbin, *The Churches Militant* (New Haven, 1973); on the Mexican War, Clayton Sumner Ellsworth, "American Churches and the Mexican War," *American Historical Review*, XLV (1940), 301-26; and the Spanish-American War, correcting earlier accounts, Winthrop S. Hudson, "Protestant Clergy Debate the Nation's Vocation, 1898-1899," *Church History*, XLII (1973), 110-18.

28. Wilson, Jackson Day Address, Indianapolis, January 8, 1915, in *Messages and Papers of the Presidents*, vol. XV, 8034, quoted in Kistler, *This Nation*, 191.

29. Wilson, War Message, April 2, 1917, in *Messages and Papers of the Presidents*, vol. XVII, 8233.

30. I have argued in *The End of American Innocence* (New York, 1959) that the First World War served as a catalyst, speeding up and altering currents of change that were already visible before the war. The breakup of the national religious consensus, already inevitable, was speeded up and brought out in the open by the war and its immediate aftermath. This is confirmed by two excellent recent books, each dealing with a major part of American religion henceforth separated from the rest and internally fragmented as well: William

R. Hutchison, *The Modernist Impulse in American Protestantism* (Cambridge, Mass., 1976); and George M. Marsden, *Fundamentalism and American Culture: The Shaping of Twentieth Century Evangelicalism, 1875-1925* (New York, 1980).

31. The standard compilation of extreme clerical chauvinism in World War I is Ray Abrams, *Preachers Present Arms* (New York, 1933). This is impressively criticized in Hutchison, *The Modernist Impulse in American Protestantism,* 232-44.

32. Eldon G. Ernst, *Moment of Truth for Protestant America: Interchurch Campaigns Following World War One* (Missoula, Montana: American Academy of Religion Dissertation Series, Number 3, 1974), x, 58, 151.

33. Robert T. Handy, "The American Religious Depression, 1925-1935," *Church History* XXIX (1960), 3-16. A recent historian of Fundamentalism argues, interestingly, that instead of a decline in religion what this period saw was a shift in power from liberals to conservatives. Joel Carpenter, "Fundamentalist Institutions and the Rise of Evangelical Protestantism," *Church History,* LXIX (1980), 62-75.

34. The beginnings for a study of the churches and World War II can be found in Nelson R. Burr, *A Critical Bibliography of Religion in America* (2 vols., Princeton, 1961), II, 633-37. I am not aware of a systematic study of church opinion in relation to Korea. An excellent beginning on the churches and Vietnam is Richard John Neuhaus, "The War, the Churches, and Civil Religion," *The Annals of the American Academy of Political and Social Science,* 387 (January 1970), 128-40.

35. One highly interesting discussion of this question is Laurence R. Veysey, "Continuity and Decline in American Religion since 1900" (paper delivered in San Francisco at the 1980 meeting of the Organization of American Historians).

36. This point is powerfully made by Anthony Wallace, *Religion: An Anthropological View* (New York, 1966), 266.

. 10 .

Europe and the
American Mind

In 1981 I was invited to deliver a paper at the Cool
Seminar in American Studies at Sapporo in Hokkaido,
the northern island of Japan (so named to distinguish
it from a seminar annually meeting in Kyoto in hot
weather). The assigned topic, "Europe and the American Mind," seemed to me at first both hackneyed and
too large, but I decided to accept and do my best with
it. Actually the topic became immensely fruitful when
considered in a Japanese context. As a series of brilliant
Japanese comments made clear, both Europe and America have been crucially important in forming the modern Japanese mind. Whether these two powerful foreign
civilizations were the same or separate, and whether each
was benign or destructive in its effect on Japan, have
been central topics for Japanese critical analysis.

I have revised the paper slightly, mainly to remove
references to the other two papers on the same subject
presented at the Seminar. The paper has been published
in Japanese translation by the Cool Seminar, and will
be published shortly in the *History of European Ideas*.

The topic we have been assigned, "Europe and the American Mind," is obviously an impossible one to treat adequately in a single lecture of bearable length.[1] It involves nothing less—in fact probably more—than an analysis of the whole nature and history of American high culture. So that you may sympathize with our predicament, I ask you to imagine a Japanese scholar asked to explain, in a single lecture, the relation of Chinese to Japanese culture, before an audience of American Far Eastern experts.

One advantage of dealing with an impossible topic is that one can say almost anything one likes. The only method possible is that of impressionism, a method most historians find far more congenial than they usually dare to admit. As long as it is understood that one cannot *cover* this topic, a glance at it should help to raise many important questions, and raising questions will be my whole purpose.

In approaching our vast topic, one almost necessarily starts with literature. From the beginnings of American writing, a central question endlessly debated is whether a separate national literature exists, should exist, or can exist—a question on which English literature departments in our universities are still not entirely agreed. One could address the question almost as well in terms of any part of high culture: painting, architecture, philosophy, or, not least, the writing of history. Relations with Europe, or the lack thereof, are the central content in the study of foreign policy during most of our history. In the vast and thriving field of social history, attitudes toward Europe have been central for the study of immigration. For the successive waves of immigrants Europe has represented what they came to escape: feudal oppression, the military draft, the pogrom. Europe has also represented the comfort of ancient ways and customs, to which, according to Hansen's law, the third generation wants to return.[2] One could also discuss our huge question in terms of the history of religion, a central field for the understanding of American culture. From the Puritans to the Roman

Catholics or the Jews of Eastern Europe, each religious group that has come from Europe to America has had to adapt its teaching and structure to American conditions, usually losing something in intensity and doctrinal purity, and gaining something in popularity, adaptability, and competitive survival.

Because of the nature of most of my training and because of the implications of the word "Mind" in our assigned title, I am going to approach the topic in the manner of traditional intellectual history. I take this to mean the history of the articulate. In America as in Europe this field has recently come under attack as "elitist." Obviously this attack is partly justified. We have not always recalled clearly that the history of the articulate is the history of those who have had the leisure and training to be articulate, and does not include the history of those who have spent their lives in a struggle to survive. This fact being made clear, there is nothing illegitimate in the study of the articulate minority.[3] Much of the interest in traditional intellectual history lies precisely in the relation of the articulate spokesmen to the less articulate masses. How much of America was actually represented by the great spokesmen of democracy—for instance Jefferson, Emerson, or William James? All these were passionate democrats; all came from the highly literate small minority.

Relying on articulate spokesmen, and for the most part leaving open the fascinating question of their representative quality, I want to direct my brief discussion to the very center of traditional intellectual history. I take that to be the point at which literary culture comes together with political thought in the broadest sense.

Let me begin by accepting a traditional dichotomy. The majority American tradition sees America as culturally independent of Europe and superior, especially in morality. The minority sees America as a somewhat unsuccessful offshoot, inferior especially in artistic culture. Nearly all our most interesting writers have found themselves somewhere in the middle of this division, try-

ing to come to terms with both sides. For many political spokesmen, the chief meaning of America has been the rejection of
Europe—of European aristocracy, cynicism, and war. For a minority the only hope for American survival has been to recognize
American membership in—recently even leadership of—the Atlantic world.

Now let us try, before starting a whirlwind chronological jour-
ney, to look at both the majority and minority view a bit more
intensely.

The first side, the majority view, is the easier to state. Democracy, morality, and progress are its elements. America is better
because it is less closely linked to the cruel and oppressive past,
more promising in terms of the broad democratic future. This
major American orthodoxy has been best defined by the two best
European critics of America, Alexis de Tocqueville and George
Santayana, both quintessential representatives of traditional Europe. For Tocqueville, the essence of American culture was the
universal belief in equality. To this European aristocrat, as to
many nineteenth-century Americans, American equality foreshadowed the destiny of Europe as well. While most Americans
believed this was messianic fervor, Tocqueville balanced the gains
and losses of egalitarianism with exquisite detachment. American
democracy offered the greatest hope for political and social stability, and the greatest dangers of intolerant mediocrity.

Santayana, whose America was mainly Boston, saw the center
of American tradition in moral idealism. American orthodoxy in
his time "consisted in holding that the universe exists and is governed for the sake of man or of the human spirit." Believing this
opinion entirely mistaken, he was able to treat its great exponents, Emerson and James, with respectful and affectionate irony.
American idealism, at least outside the campuses, was no foggy
German belief that the highest reality was unseen. The American was "an idealist working on matter." The goodness of the
universe was demonstrated every day by practical improvement
and the solution of mundane problems. Santayana could treat

American practical idealism with considerable respect, yet he never found it congenial. The moral optimism on which it was based precluded, he said, serious poetry and serious religion.[4] Whether young Americans respond favorably or unfavorably to a first reading of these two great critics serves as a clear indication whether they belong on one side or the other of the ideological divide about the relation of America to Europe. To some, both come as a startling reinforcement of their own inner doubts. To most both are hard to understand: questioning the value either of democracy or optimism seems almost incomprehensible.

The minority view, the belief that America is inferior to Europe, is harder to analyze. Here too Santayana especially makes a good starting point:

> The luckless American who is born a conservative, or who is drawn to poetic subtlety, pious retreats, or gay passions, nevertheless has the categorical excellence of work, growth, enterprise, reform, and prosperity dinned into his ears; every door is open in this direction and shut in the other; so that he either folds up his heart and withers in a corner . . . or else he flies to Oxford or Florence or Montmartre to save his soul—or perhaps not to save it.[5]

It is true enough that real conservatives are rare and discontented in the United States. Real conservatives believe that it is best for people to accept the conditions in which they live, because attempts to improve things will probably make them worse. Our recent history shows that a politician who even begins to question the ability of the United States to advance rapidly on all fronts at once soon gets into trouble. The big winners—Franklin Roosevelt, Dwight Eisenhower, Ronald Reagan—are those who can say that there is nothing to fear in the modern world but fear itself; that unity and resolution will give us the peace, prosperity, and freedom that have always been our proper destiny; that with enough will-power we can cut taxes and increase defense spending at the same time.

Yet notoriously, real radicals in the United States have been almost as rare, and almost as unhappy when they have turned up, as real conservatives. To believe that the existing system must be destroyed root and branch, even at the cost of strife and suffering, is as incompatible with American practical idealism as to say that nothing can really be changed at all. Perhaps what the many generations of American dissenters and expatriates have looked for in Europe is extreme individual commitment of any kind. Whether radicals or conservatives, they have wanted to escape from the persistent pressures of compromise, moralism, progressivism, and optimism. They have wanted to think more deeply and to speak more frankly than has usually been possible in America. (As Tocqueville saw so clearly, the chief restraint on free speech in America has not lain in repressive laws, but in the sheer power of moral consensus.) Exiles and expatriates have wanted to think the unthinkable, to say the outrageous, to push things to extremes, whether of frivolity or despair, whether of anarchism or socialism, existentialism or estheticism. Usually this sort of extremism, whether of revolution or reaction or sheer individualist irresponsibility, has had something aristocratic about it. It involves a deep belief in the supreme importance of the cultivation and expression of the individual spirit, for its own sake, without regard for public consequences. For this American dissenters have found sanction in the values—ultimately aristocratic values—exemplified in European culture.

To reject practical idealism, to question morality or progress, can be tremendously liberating for Americans abroad. Lambert Strether, the principal "ambassador" of James's great novel, gradually loses touch with the moralism of Woollett, Massachusetts, and finds himself condoning the hedonistic ways of Paris. As he does so, he feels suddenly *younger*. On a far lower level, a somewhat similar kind of liberation is sought in the same novel by Jim Pocock, who takes it for granted in the manner of generations of American tourists that in Paris one should taste the illicit excitements of the flesh. Plenty of Americans of impeccable moral rigidity at home have felt it almost a duty, abroad, to sample at least the Folies Bergère. In a not entirely different manner, nineteenth-century Protestants like Harriet Beecher Stowe experienced an almost guilty *frisson* when they found themselves, to their surprise, responding emotionally to a great Catholic ceremony or cathedral that represented everything they had been brought up to fear and hate.

To sum up this division, for most Americans, America has meant liberation from Europe. To some Americans, Europe has meant liberation from America. And to the most interesting Americans, the choice between these kinds of liberation has been a hard one.

Since I am a historian by training, I cannot help trying to arrange in chronological order a few episodes and examples of this fundamental division in American views of Europe. One might conceivably start at the very beginning, when most of the first settlers must have felt—and some expressed—all sorts of mixed hopes and qualms as their ships pulled away from familiar European shores. I will begin, however, in the late eighteenth century, when American cultural as well as political independence started to take shape.

Right before the Revolution some upper-class Americans were trying hard to achieve in America a fully European way of life. To some this meant more luxury in clothes or houses and more attention to the niceties of aristocratic behavior. To others it meant struggling for a better support of scientific achievement and literary culture. At the same time, however, a larger part of the people were turning increasingly against Europe. The preachers of the Great Awakening in religious terms and the Commonwealth pamphleteers in political terms both insisted that Europe and especially England was luxurious, irreligious, decadent, and corrupt. America's duty was to remain a place apart, dedicated to pure religion and morality, to frugal ways of life and simple forms of government.[6] Believers in these two views were forced to make a choice—for those in the middle often an agonizing choice—in the time of the Revolution. The majority opted, on various grounds, for secession from Europe and a new beginning.

For a few, the choice was easy. Thomas Paine, an Englishman, represented one extreme view—that America must be as different as possible from Europe: "We have it in our power to begin the world over again."[7] Alexander Hamilton represents the other extreme. Hamilton, alone among the major figures of the age of Revolution, wanted America to become a great nation on the European model: rich, powerful, cultivated, urban, and armed. To achieve this Hamilton believed, following such European masters as Machiavelli and Hume, that it was necessary to mobilize the European passions for wealth, prestige, and especially glory. It is interesting that Hamilton, unlike most of the other great figures, never visited the old world. Somehow, as his great admirer Talleyrand pointed out, "il avait deviné l'Europe."[8] But for most of the great figures of the period, partly European in culture but American in choice of allegiance, the choice was less clear and easy. Benjamin Franklin, who superbly impersonated a rustic American, actually relished the brilliance and diversity of European society, and was completely at home in both London and Paris. John Adams, insistently provincial and puritanical, was uneasy in both capitals, and said in a homesick moment in London that when he went home he would miss only the bookstores, conversation with four or five London acquaintances, and his correspondence with Thomas Jefferson in Paris.[9]

Jefferson, the greatest spokesman of American political independence, exhibits in his life and writings the deepest ambiguities about Europe. On the one hand, he never tires of warning his countrymen against European vice, cynicism, and luxury. A European education for a young American is worthless and debilitating; America must learn from European precedents to avoid the moral dangers of urban development. Yet Jefferson in Europe delighted immeasurably in European books, architecture, music, and intellectual conversation. He exhausted his fortune and com-

promised his antislavery principles in a lifelong effort to reproduce, in the Virginia piedmont, a way of life that included a
mansion built on rigorously correct European lines, equipped
with a full supply of European books, pictures, statues, and
wines. He imported a French maître d'hôtel and once hoped, in
vain, to import an orchestra.[10]

In the immediately post-revolutionary generation confidence
in American superiority reached new peaks. According to some
patriots America, already far ahead of Europe in freedom and
well-being, was on the verge of passing her also in science, painting, and especially in literature, whose American triumphs would
be expressions of republican purity. Unfortunately support for
science and the arts was hard to come by, and new republican
genres failed to appear. A new kind of hostility toward Europe
came into being among conservative Americans during the later
stages of the French Revolution. Europe was not only decadent,
she was also a dangerous exporter of revolution. While many
Americans refused to be alarmed, only a very few became close
followers of the French program for world revolution. In the
standard American view of history, the orderly and moderate
American Revolution was henceforth contrasted with European
revolutionary violence and irreligion, produced by European oppression and popular ignorance. Only in 1815, with the failure
of British invasion and the stalemate peace of Ghent, were Americans able to withdraw completely from European politics, free
from fear and confident that they owed their freedom to their
superior institutions. In 1823 these attitudes were promulgated
in a major official statement. The Monroe Doctrine formalized
American belief that the European political system was "essentially different" and had no place in the entire New World.

The nineteenth century, which in American historiography
runs from 1815 to 1917, is the most important century for historians of American culture. It is in this period that the United
States had the opportunity, unparalleled among major modern
nations, to work out her own political, economic, and social des-

tiny without serious fear of foreign interference. If there *is* such a thing as a separate American culture, it is in this period that its early development must be found. It is America in this period that is characterized by Tocqueville and Santayana. When literary "redskins" fight with "palefaces" over the question whether American literature should be native and spontaneous or polished and European, they usually cite on both sides the great nineteenth-century American authors, starting with Whitman and James.[11]

It is well worth pausing a moment to think about this whole century of relative political and cultural separation, as important perhaps for America as the Tokugawa isolation for Japan. It is also helpful, and conventional among American historians, to divide it into three periods, usually seen as the period of Jacksonian Democracy, the period of industrialization, and the Progressive Era.

In the early nineteenth century democracy, for Americans, stopped being one of the three classic kinds of government with its own virtues and faults, as it had been in the European tradition since Aristotle, and became instead a special revelation. This belief was dominant and very powerful, whether or not Andrew Jackson had much to do with it and whether or not American claims of real equality were exaggerated. The special democratic revelation delivered to the United States was inseparably mixed with Christianity in its popular American forms. In the long run, the Republican millennium was the manifest destiny of the whole world. This was evidenced by the more orderly, nationalist, and constitutional of the European revolutions.

At its extremes, Jacksonian hostility to European culture could sound like opposition to culture itself. This was the case in the attack of Jacksonian democrats on John Quincy Adams, an American patriot who had been educated in Europe by his father and Thomas Jefferson, and who hoped to create in the United States something like an Enlightened European nation complete with a national university and official support for the arts and sciences. His successful opponent was hailed as an untutored American genius fresh from the woods.[12] From here on it was advisable for American statesmen to look hard in their backgrounds for a log cabin, and at least to avoid any hint of a European mansion.

In this period of continental expansion the dominant democratic patriotism could be pushed into extremes of demagogy and chauvinism, yet its genuine commitment to equality gave it great moral power. Democracy, progress, and superior morality were a hard credo to resist. Serious dissenters, as Tocqueville saw, needed hardy souls.

The most common variation to the dominant anti-European mood was that which made England an exception, and this was common only in the older parts of the country. In New England early Victorian Britain, increasingly evangelical and moral and even developing slowly toward democracy, often seemed to be a partner with the United States in opposing the immorality, political backwardness, and violent revolutionary potential of continental Europe. This was the belief of upper-class Boston during the great period of New England literature. Some critics, poets, and historians, like Ambassador James Russell Lowell, were as much at home and as much admired in London as in Boston. In the Southeast some hankered after another England, conservative and aristocratic, and the affection was to be reciprocated by the governing British aristocrats who sided with the South in the Civil War.

Of the major writers of the period, only Whitman sided wholeheartedly with indigenous culture, and only Poe longed unreservedly for the holy ground of classic Europe. The rest—Emerson, Hawthorne, and Melville—were all deeply involved with America's democratic hope and conscious of European cultural depth, and each had to work out his own difficult balance.

In the late nineteenth century, America, a little before Japan, followed the major European nations into industrial development. She also attracted an immense new European population.

In spite of these facts, most Americans denied any close similarity between America and Europe. Until almost the end of the century, the main note sounded by spokesmen of foreign policy was isolationist. From time to time minor incidents produced brief and unconvincing threats of war, usually with England. On the other hand some of the American rich avidly took up the literal imitation of European architecture and English private schools, and became indefatigable collectors of European art. Some of the same people who admired European artistic genius were increasingly fearful of European immigrants and European social theories.

By the end of the century a group of upper-class intellectuals appeared who were secure enough in their own, usually inherited position and familiar enough with Europe to avoid the conventional extremes of isolation and imitation. Henry Adams, often irritated with England and consistently hostile to France and Germany, ambivalent about American democracy in the manner of his New England ancestors, worried during his long life about the common doom of both branches of Western culture. And no serious critic reads Henry James as either uncritical of Europe or dismissive of America.

Admiral Mahan, one of the first American military intellectuals on the European model, dismissed pacifist dreams and wanted America to ensure her own safety and greatness in the ways indicated by the history of Venice, Spain, and Britain. From time to time Mahan's friend Theodore Roosevelt, a baffling mixture of European culture and American bounce, hankered like Alexander Hamilton (whom he admired) for European kinds of power.

Roosevelt and his circle played some part in the coming of the Spanish-American War, traditionally taken as the point at which America entered European great-power politics, surprisingly enough in the Asian theater. It was a surprise for most of the people when the new question of American overseas empire emerged out of a traditional intervention against Spanish monarchical oppression in the Western hemisphere. The debate over empire was in part a debate over the Europeanizing of the United States, with the anti-imperialists insisting that a rule over subject peoples would violate sacred traditions of American moral superiority. President McKinley finally settled the question by announcing that America would take the Philippines not for realistic European reasons of defense or profit but as a part of the age-old American mission of spreading democracy and Protestantism. In this, it is worth recording that he was supported by only a part of the Protestant clergy[13] and opposed by William Jennings Bryan, a perfect representative of traditional America at its best and worst, and a statesman who remained profoundly suspicious of Europe throughout his long career.

The acquisition of a Pacific Empire did not, as its opponents had predicted, immediately or obviously move America along the path of declining imperial Rome. Most Americans in the Progressive Era continued to believe in American isolation and moral superiority, and indeed a distinctly anti-European note was struck by many progressives. This was especially true in the Western branch of Progressivism, which had Populist roots and tended to see both Europe and the Eastern states as the homes of snobs, bankers, and militarists. An anti-European version of American history, first proclaimed by Frederick Jackson Turner in 1893, swept to victory in these years. Most of the rising generation of influential American historians began to teach that American democracy and individualism, the distinguishing features of our civilization, came from the frontier and had nothing to do with such European imports as Puritanism or the Enlightenment. The European model for American literature was challenged by realistic Western novelists and by prairie poets with their roots in Whitman.

Against what seemed to them a tide of Western barbarism the embattled defenders of European culture stuck to their fortresses in the Eastern colleges and periodicals. At Harvard especially,

Professor Irving Babbitt insisted with learning and clarity that the only hope for civilization lay in maintaining the classical standards that had prevailed in Europe before the French Revolution.

The ancient argument about America's relation to Europe was sharply intensified by the outbreak of war in 1914. To Midwestern progressives, and indeed in the first war years to most Americans, America's duty to herself and to the world seemed to be to preserve an island of peace and progress. On the other hand from the start the Eastern apostles of European culture believed that it was America's duty to save civilization by joining the heroic struggle of England and France.

The decision was made by a man who combined elements of both sides. Woodrow Wilson loved traditional English literature. In his arduous climb up the academic ladder he had several times, at seasons of nervous exhaustion, sought inner peace in visits to Wordsworth's Lake Country. Yet he was also an epitome of American moralism, with a fierce Calvinistic contempt for European corruption, militarism, and balance-of-power politics. When he finally led America into war, Wilson appealed to our highest moral traditions and called for an immense commitment to make Europe over in the image of American democracy. When Revolution came in Russia he and some of his advisers believed that this vast mysterious country was at last moving in the same direction.

Wilson's misinterpretation of Europe was part of the reason for his and America's tragic failure in 1918-20. The cause and result of this failure was the harshest repudiation of Europe in American history. In the postwar regime the Midwest, which had never been convinced by Wilsonian internationalist rhetoric, gained new power. The most popular organs of public opinion depicted Europe as hopelessly decadent and deadbeat, with leanings toward bolshevism.

Throughout the 1920s, when this anti-European stereotype was dominant in the Congress and the popular press, the New Humanists, heirs of the genteel tradition, bravely defended the old

Europe of rigorous literary and intellectual standards, a Europe hard to locate in the present world. A very different version of European superiority, and one far more powerful in this period, was proclaimed by the rebellious literary intellectuals.

The vision of Europe seen by this diverse and famous group was not without its contradictions. To some American literary rebels Europe was a land of wine, leisure, ripe tolerance, and love of art—a refuge from what Van Wyck Brooks called Puritanism or what H. L. Mencken called "the uplift." For others Europe was a source of strenuous, extreme, and bold innovation in literature and the arts. For some American expatriates, their love of Europe was not deep enough to survive an experience of European reality, perhaps in the person of a French landlady. For others—for instance for T. S. Eliot, who had learned his Europe partly from Babbitt and Santayana, allegiance to European tradition was a deep intellectual and moral commitment. For some of the most talented expatriates, as for Henry James in an earlier generation, the choice of Europe as against America was not at all simple or one-sided. A sojourn in Europe and an immersion in modern European culture helped some Americans to discover the resources in the American tradition. It was in the twenties that Melville, Hawthorne, and Mark Twain were rescued from the traditional moralistic and progressive simplifications of American literary history.

The 1930s brought a repudiation of Europe by many kinds of Americans. Confronted with a disastrous depression, most Americans were absorbed in their own problems, and in the early New Deal period Roosevelt's economic nationalism replaced Hoover's relative internationalism. Congressional committees blamed the Great War on munition makers, and American participation on British propaganda. Pacifists, isolationists, and students swore that they would never again be sucked into the quarrels of Europe.

As the expatriates sailed home, literary intellectuals sometimes put together an odd combination of cultural patriotism with

shallow Marxism.[14] Some American intellectuals replaced an uncritical admiration of Western Europe with a still more uncritical glorification of the Soviet Union. The Communist Party,
however, reaching the high point of its influence in the period
of the Popular Front, made heroes of Jefferson and Lincoln and
insisted that Communism was "twentieth century Americanism."
Sometimes under radical auspices and sometimes not, American
writers and artists discovered folk art; American folk music was
studied, collected, and performed; and first-rate regional novelists concentrated on the heritage and mores of the South and
West. It is no coincidence that this period saw the beginning of
the American studies movement in the universities, and the
movement has reflected some of the period's flavor.

With the approach of a new World War a great debate over
American participation recapitulated in part the debate of 1914-
17. The movement to aid England, backed once more by Eastern
intellectuals, was powerfully resisted by isolationists of the left
and right, some of whom argued that Europe was hopelessly
doomed and should be written off. To swing the majority de-
cidedly toward war, even against Hitler, required actual foreign
attack.

The immense American intervention of the 1940s in both Europe and Asia put an end for a long time to any belief in European superiority. The isolationists of the right, forced by events
to support the war but never forgiving Roosevelt, showed their
feelings by insisting that Asia, not Europe, was the really important theater. Some conservatives were deeply uneasy about the
Russian alliance, while liberals were equally suspicious of a revival of traditional European imperialism. The soldiers, as soldiers often do, tended to dislike their wartime allies and to prefer, when they encountered them, the people of the enemy nations.
To a great many intellectuals and others, it became difficult to
believe in the cultural superiority of a continent that, except for
England, had succumbed to a deeply anti-cultural totalitarianism. For a while, the Europe in which Americans had long be-

lieved, whether as the haven of culture or the source of newness, was not there.

The 1950s, the age of American political and cultural empire, badly needs to be looked at and understood by American cultural historians. In the period of the Marshall Plan and then NATO, America forged far closer ties with Europe than had ever existed before. This was accomplished against the bitter opposition of various and formidable anti-Europeans running from Charles A. Beard to Senator Robert Taft. Dean Acheson, the symbol of unity with Europe against the communists, was the bizarre choice of Senator Joseph McCarthy as the central target for his anti-communist crusade.

In this period many intellectuals for many reasons found themselves less alienated from America and less respectful of Europe than ever before. In some cases, this tendency resulted from a discovery that Communism as well as Fascism was hostile to intellectual creativity, and a renewed interest in American traditions of pragmatic compromise. Some believed, mistakenly as it turned out, that America through the New Deal and Keynesian economics had permanently solved the problems of poverty. Many Americans and some Europeans believed that world cultural leadership had passed to the United States along with world power. In the 1950s America was clearly the center of excitement in painting, and arguably in the lead in the natural and social sciences. Support for high culture seemed to be growing. Like many others, Jacques Barzun, a humanist of European origin, came to believe that now "it was Europe that was provincial."[15]

In some cases the new interest in America and respect for American achievement was a healthy corrective to earlier habits of ignorant self-denigration and romantic idealization of Europe. But the new patriotism of intellectuals could also lead to intellectual chauvinism and smugness. Some Americans, perhaps most effectively Reinhold Niebuhr, warned America against arrogance and complacency, but not many of these pointed to Europe as an example of measure and modesty. One of the few groups to argue that America should follow European leaders was the New Conservatives, who tried unsuccessfully to convince their fellow-citizens of the greatness and relevance of the tradition of Edmund Burke.

In the 1960s, as none of us can forget, American complacency was suddenly and dramatically challenged by domestic racial crisis, a revolution of youth, and a foreign war which grew ever bloodier, less successful, and more unpopular. Yet in the revolts of this period, few looked for guidance to the prosperous and Americanized Europe of the day. Especially in the student movement, materialism was the main enemy, and Whitman and Thoreau seemed to offer more help than Marx and Lenin. In the widespread revolt against modern mass culture, the influence this time ran from America to Europe rather than the other way. European student grafitti used American obscenities, and the dress and music of European youth were largely American in origin.

It is very hard to look at the immediate present in the perspective of such a rapid survey. It would be tempting to say that ever since World War II the old habit of contrasting America to Europe has been obsolete. Certainly Americans can and do go to Europe in large numbers without encountering much serious difference, whether they travel in the convention of the Hilton or the hostel, with matched luggage or backpack and guitar. In Europe today it is not easy to find the serene aristocratic ripeness loved by James, and America hardly seems now, as it did to him, the home of naïve moral innocence. Few could say today, as William Dean Howells did in 1886, that Russian literature could hold little interest for Americans because here the smiling aspects of life were dominant and the sum of hardship and oppression small.[16] America is still a country of comparative freedom, but in large sectors there is enough violence, misery, and despair to fuel plenty of tragic novelists.

Yet the old and fruitful relation may not have changed quite beyond recognition. In the 1970s, some Americans still thought like Gertrude Stein in her day that Paris was where the twentieth century was. The radical and uncompromising thinkers of many kinds that our graduate students follow as intellectual leaders mostly live in Paris, sheltered by the triumphantly elitist educational system built by Richelieu, Louis XIV, and Napoleon. On the other hand, today as in the past some are invoking European wisdom and moderation in foreign affairs to hold in check a new and dangerous form of American political innocence. Our subject, after all, has been Europe and the American *mind*. Perhaps, for the American mind, Europe has always consisted mainly of images formed to meet American needs.

Notes

1. Cushing Strout, *The American Image of the Old World* (New York, 1963), is an admirable attempt to deal with this whole subject at book length. Philip Rahv. *Discovery of Europe* (Boston, 1947), is a well-chosen anthology of American reactions to Europe with intelligent editorial comments.

2. Marcus L. Hansen, *The Problem of the Third Generation Immigrant* (Rock Island, Ill., 1938). This idea is developed in terms of religious history by Will Herberg in his *Protestant, Catholic, Jew* (Garden City, N.Y., 1955).

3. This and related issues are discussed in several of the essays in *New Directions in American Intellectual History*. John Higham and Paul K. Conkin, eds. (Baltimore, 1979).

4. George Santayana, *Character and Opinion in the United States* (New York, 1920; paperback edition, 1956), 11, 108; "The Genteel Tradition in American Philosophy," in *Winds of Doctrine* (New York, 1913; paperback edition, 1957).

5. *Character and Opinion,* 105.

6. J. G. A. Pocock distinguishes most suggestively between the "party of virtue" and the "party of commerce" in the eighteenth century and, by implication, thereafter. "Virtue and Commerce in the Eighteenth Century" (review article), *Journal of Interdisciplinary History,* III (1972), 119-34.

7. Paine, *Common Sense and Other Writings* (reprinted Indianapolis, 1953), 51.

8. Quoted in *The Letters and Journals of George Ticknor* (Boston, 2 vols., 1876), I, 261.

9. Adams to Jefferson, March 1, 1787, in *The Adams-Jefferson Letters*, Lester J. Cappon, ed., 2 vols., Chapel Hill, 1959), I, 177.

10. See Eleanor D. Berman, *Thomas Jefferson among the Arts* (New York, 1947), 175-76.

11. These terms were coined by Philip Rahv, "Paleface and Redskin," in *Image and Idea* (New York, 1949), 1-5. This celebrated essay carries forward a distinction suggested earlier by Santayana and developed further by Van Wyck Brooks and others.

12. John William Ward, *Andrew Jackson: Symbol for an Age* (New York, 1955).

13. See Winthrop H. Hudson, "Protestant Clergy Debate the Nation's Vocation, 1898-1899," *Church History,* XLII (1973), 110-18. This article corrects earlier accounts.

14. See Charles C. Alexander, *Nationalism in American Thought, 1930-1945* (Chicago, 1969).

15. Barzun in *America and the Intellectuals, A Symposium* (New York, 1953; originally published in *Partisan Review,* in 1952). This whole symposium expresses brilliantly the intellectual turn toward America and away from Europe.

16. Howells, *Criticism and Fiction* (New York, 1891), 128.

Review Articles

· I ·

Perry Miller's Parrington

During my graduate years at Harvard (1937-41 and again 1945-47) I was not one of Perry Miller's own doctoral students nor was I ever one of his intimates. However I did take courses with him and also saw him in action in the political controversies at Harvard in 1939-40, when he was at the height of his intellectual and polemical powers. My two first books were reviewed by him, one favorably and one, I thought, harshly. I met Miller again in 1963-64 when I was spending the year at Harvard beginning to study the Enlightenment. When I complained about his treatment of my book he characteristically invited me to lunch and talked expansively about his current immense project, the subject of this review. A few weeks after this lunch I read of his death and was moved to write this article as a tribute. Perry Miller's flaws have been exhaustively examined by a generation of critics. The flaws are real, and this was a necessary task. Yet Miller remains the most powerful teacher and writer of history that I have encountered.

Miller's *Life of the Mind in America* was published

by Harcourt, Brace and World in 1966. This review article was published in *The American Scholar,* XXV, 3 (Summer 1966). The editors have kindly transferred the copyright to me.

The late Perry Miller, according to his intimates, sometimes referred to his long-awaited, long-delayed work on American thought in the nineteenth century as "Miller's Parrington." The most obvious meaning of this phrase was simply that the projected work, like Vernon Parrington's *Main Currents of American Thought,* was to be a vast piece of literary architecture. Indeed, Miller's work, if completed, would have dwarfed Parrington's. Where Parrington had concentrated on literature and politics, Miller planned to deal with a whole range of topics such as science, religion, law, education and philosophy. A generation of intensive scholarship since Parrington had immeasurably extended the material confronting a historian. In the 1960's, moreover, Parrington's clear dichotomies between East and West, Conservative and Liberal, were unusable for any serious historian, let alone for so complex and subtle a critic of American culture as Perry Miller. One cannot help wondering whether the phrase may not have implied that Miller, like Parrington, would fail to complete his magnum opus. The insight, the flamboyance and the bitterness of such a meaning would be entirely characteristic of the man.

As it turned out, while Parrington died halfway through his third and last volume, Miller left behind only two completed books out of nine, with a fragment of one more. Often posthumous publication of uncompleted fragments is a mistake; in this case it was not. In some ways *The Life of the Mind in America,* although it is by no means Miller's masterpiece, is the most fitting monument this powerful and fascinating historian could have.

Miller, who was a prolific and thus necessarily an efficient writer, bogged down in this book well before his death. Whatever the personal reasons for this fact, there are sufficient explanations in the nature of the task he had set himself. His greatest book, massive in scholarship and ample in scale, dealt only with New England, and almost entirely with Massachusetts, from the mid-seventeenth century to about 1730. His last project proposed to deal with the whole amorphous, expanding republic from the Revolution to the Civil War.

The subject is not only a big one, but it is also one peculiarly formidable for the intellectual historian. No clear, articulated system of thought stood at the center of the national enterprise, as Puritan theology had stood at the beginning of New England. Theology itself was still a lively and controversial discipline in the seminaries, but most Americans ignored it. The combination of Scottish philosophy, English political economy and attenuated Calvinism taught in most colleges inspired few of the period's major achievements. Politics was the most obvious axis, but did not really subsume the entire Life of the Mind, and Miller's interest in political history had always been secondary and indirect. The lack of a central theme probably dictated Miller's decision to subordinate chronology, and organize his work into nine vast topical surveys, each covering the entire period. This decision, like most decisions of historians to avoid movement through time, was probably a mistake.

The flow of time, in Miller's best books, had been channeled into stories of decline. Yet to treat this period of the expansion of the Republic and the emergence of a distinctive democratic culture solely in terms of decline would have been grotesque. In some terms, as Miller shows us, decline indeed took place. James Kent and Joseph Story, the great codifiers of American law, are as close as anybody to being heroes of the completed sections of the book. To post-Federalist gentlemen like them, their period was one in which cosmopolitan culture and intellectual discipline were increasingly threatened by ignorance and demagogy. In terms of political thought, the period from Jefferson and Adams to Jackson and Webster may be seen as decline, and in terms of political institutions the story goes from the success of 1787 to the breakdown of 1861. Both the completed sections of the book, that on religion and that on law, are in part stories of vulgarization. Yet the 1850's are the period of the greatest triumphs of American literature, and the emergence from nowhere of Lincoln makes it hard to speak of exhaustion even in politics.

If, moreover, Miller had centered his whole story on decline, it would have had to be decline from the Enlightenment. Miller never felt easy with the Enlightenment, and much of his best work is a polemic against its assumptions. His long list of books skips it, jumping again and again from the Puritans to the Romantics and Transcendentalists. In this book, as in some of his others, its ideas are treated only in passing, and this is one of the book's major weaknesses. Nobody yet has studied the American Enlightenment, transcending politics, in wide scope and analytic depth. One would think that Miller might have enjoyed a harder look at its glorious paradoxes, its planets spinning endlessly in accord with the dictates of moral law. Perhaps because of the sheer uncongeniality of the only possible starting place, Miller provided no firm pad from which to launch his story.

If decline could not furnish the main movement of the book, neither could advance. Toward all the possible themes of triumph, Miller was deeply and even agonizingly ambivalent. It is this fact that gives the book its originality and interest, and it is very likely the same fact that made it impossible to complete. Since Turner, indeed since Bancroft, most historians have treated the early nineteenth century as a time of expansion—of the frontier, of industry, of the national ego. In a recent review in the *New York Review of Books,* Alfred Kazin, usually a most perceptive critic, seems to say that Miller identified himself completely—despite some foreboding asides—with American power and might. This I think mistaken, in emphasis if not in its basic assertion. Miller did indeed sound at times the Hemingwayish note of a he-man historian among the literary sissies—his most damaging foible. He talked a good deal, and with relish, about his wartime experience, and referred to his early exposure to the jolly brutalities of the I.W.W. In a flamboyant late preface mentioned by Kazin, he compared his own realization of American force, while helping load oil drums on a river steamer on the Congo, to the creative vision granted Gibbon while sitting in the ruins of the Capitol at Rome. It is worth noting, however, that the historian to whom Miller chooses to compare himself did not write about the triumph of the Roman Republic.

Miller's imagination was indeed grasped, as what historian's is not, by the awesome fact of American power. But his attitude toward this power, and toward the triumphs of the machine in general, was no more one of cheerful acceptance than that of Henry Adams. In an essay published in *The American Scholar* in 1961 Miller discussed "The Responsibility of Mind in a Civilization of Machines." Always unsympathetic to easy rejections of reality, Miller pokes some rather heavy fun at young executives who take up collecting Thoreau, and rush home in their automobiles to mix a little Walden with their martinis. Yet mechanization is presented seriously as monstrous, and abdication of responsibility before it as a species of infantile regression. The mind, Miller tells us, has given up responsibility even for destruction, which has become as automatic as anything else. Neither acceptance nor mere despair is the essay's message. It is a jeremiad, and jeremiads are preached only to the people of the Covenant, who have the duty of repentance whether or not they are able to be saved. The essay ends with a call for reassertion of the individual intellect at all costs.

In his earlier books, all Miller's heroes or even part-heroes are lonely resisters. The only person of whom he has written at length in tones of unqualified admiration is Jonathan Edwards, and Miller dwells on Edwards' rejection by the cheery Philistines of his expansive day. Edwards confronted the most profound human problems; Thoreau eluded some of them. Thus, Miller shows us, the undeniable literary achievements of the Concord ascetic are marred by his inhuman and sometimes self-protective withdrawal from experience.

In the present book, Miller did not reach the sections in which he would have had to deal with the literary figures he most admired. But he makes clear why he could not use these as the central figures of the story; he knew both them and their milieu too well:

> Because most of our classic literature of these years is hostile toward, or at least resistant to, the machine, we forget against what a background of loud hosannas Thoreau and Melville wrote.

Toward expanding power and toward individual secession Miller is ambivalent, and so is he toward democracy itself. So are many American historians, but their courage or clarity fails them when it comes to admitting it. Nearly all, in their own time, have their hearts firmly on the side of the people, and a great many have given time and money and energy toward the effort to make democracy work. Yet somehow, so far, until some future breakthrough, they find it hard to love the immediate consequences of egalitarian culture, and even sometimes to accept the outcome of egalitarian politics. The easiest way to handle this problem is to shut either one eye or both, and this was to Miller impossible. He had to look straight at the triumph of the many over the few in both its destructive and its creative aspects. None of our historians treating the age of Jackson has managed to do this; it was accomplished only by the greatest of the period's own imaginative artists.

A Menckenian dismissal of democracy was as impossible for Miller as a Whitmanesque paean. The triumph of the people over the Federalist or Virginia elite cannot be presented as an unqualified blessing by a historian who deals, for instance, with religion, education or foreign policy. Yet the democratic undertaking was far too big to laugh at; Jacksonian America believed not only that all of us, freely united, could earn our livings and govern our states, but that we could develop our own minds and save our own souls.

In America, unlike Europe, romanticism and democracy go together more often than not, and Miller's ambivalence took in both. The bland intellectualism of the Enlightenment repelled him as much as the easy gush of the feebler Transcendentalists. After Edwards, in a nineteenth-century and democratic context, nobody was able to make an intellectual system out of the mystery and terror of the Universe. In the widening gap between mind and emotion, one choice led to superficial dryness, the other—pursued with sufficient ardor—to Captain Ahab's solipsistic nightmare.

Ambivalence and ambiguity are of course the familiar stock-in-trade of modern criticism, but with Miller they go almost too deep to be usable. I suspect his ideological divisions were fundamentally a private matter, but they were also affected by the intellectual history of his own period, and demonstrated in his successive attitudes toward intellectual fashions. Miller started his scholarly career, he tells us, in the heyday of the Menckenian onslaught against standard American culture. For his subject he chose Puritanism, Mencken's favorite and easiest target. This involved him in a vindication of the past, but in a far deeper attack than Mencken's against the reigning progressive optimism. In the thirties, Miller, like other literary intellectuals, was forced to choose a political position in relation to the shifting combinations of Popular Front intellectual politics. He did not remain aloof but yet refused—sometimes awkwardly and at the cost of important friendships—to accept the dominant simplifications. After 1945, the prestige among intellectuals of neo-orthodoxy seemed suddenly to make Miller's kind of history up-to-date. Miller had long been in touch with this current, and in a phrase he attributed to Morton White, he became at least an associate member of "Atheists for Niebuhr." His work had long been full of the irony to which Niebuhr gave theological definition. It was, as Niebuhr has stated, the virtues of America that—embedded in egotism—nourished and magnified her historic faults. Yet here too Miller avoided, unlike some of the followers of fashion, the

easy way out. In a respectful review of a Niebuhr book he points
out the dangers of ironic self-indulgence:

> . . . the contention fully established, a gnawing question arises:
> by what further and extraordinary convulsions of the human
> spirit shall paradox itself be saved from becoming merely a con-
> ventional way of explaining—and thus to some degree of ration-
> alizing and justifying—the unrighteousness of the righteous.

Miller's pervasive irony was perhaps confirmed, but it was certainly not begun by his exposure to Niebuhr. His works on Puritanism all illustrate the slogan that nothing fails like success.
In the present volume, perhaps because of its closer-to-home subject or perhaps because of exhaustion, the irony is always present, often illuminating, sometimes devastating, and occasionally
forced or insistent.

The first section of *The Life of the Mind,* a magisterial survey
of the period's religious history, draws on much first-rate recent
scholarship (although often unduly harsh to other historians,
Miller cordially welcomed help). The repeated and increasingly
mechanical revivalism of the period offers an easy subject for
Menckenian caricature. It was in the Jacksonian period that
American religion, like American politics, took on its awesome
and preposterous shape. But Miller handles the ironies of revivalism with respect and restraint; they are too big to be funny.
Concentration on God led straight back to concentration on the
community; voluntarism and sectarian fission carried to an extreme led to moral conformity; the spontaneous religion of the
heart led to the development of the most cynical devices to co-
erce the emotions into acceptance of the message. Still more strikingly, an Errand into the Wilderness which had started with the
departure of a saving remnant from the European Egypt led in
this period toward the most grandiose possible return. The puri-
fication of America was seen to be merely a modest preliminary
for the permanent, rapid, systematic evangelization of the world.
Where this moral Manifest Destiny was to lead in the long run is

suggested by Miller in terms whose connotations extend beyond religious history and into the worst of our present ironies:

> One may easily ridicule this nineteenth-century rhetoric. Yet when one considers the twentieth-century role assumed by the United States, he is bound to admit this same rhetoric contains the germ of that national disposition toward "heathen" communities which would have been incomprehensible to a Jefferson or a Washington.

The second part of the book, on law, is the most finished and subtle section, and testifies to Miller's enduring energy for research in fields that were new to him. Here the lawyers, the new upper class of the young Republic, are seen fighting their heroic battles for the liberal (which means the eighteenth-century gentleman's) mind against irrationality and demagogy, against the religious geysers and the political torrent of the Jacksonian period. Clearly, the lawyers speak for the Enlightenment, and yet their main concern is a defense of the antique Gothic mystery of the common law against rude rationalization. Like starchy Presbyterians resisting revivalism, they quickly lose their grip. Originally defenders of the cosmopolitan, neutral, rational scholarship of the law, they are reduced to arguing that law embodies the truths of Americanism, morality and even, specifically, Christianity. Approaching the institutional breakdown of 1861, they are so fearful of abolitionist excesses that they defend anti-abolitionist mobs and lose all ability to save the Union to which they are devoted.

Perhaps the deepest, and certainly the most overt irony is contained in the unfinished section on science. (Since it is partly in outline, we can look at naked thematic statements which would have been embodied in metaphor and example in the finished work.) From the contemplation of the majesty of the physical universe preached by the gentlemen-philosophers of the Enlightenment, American science shifted its emphasis to crass practicality. Moreover, its spokesmen insisted that America's inventive genius, together with her free institutions and lack of moral corruption, would enable her quickly to outstrip the Old World:

> Could not this technological majesty join with the starry heavens above and the moral law within to form a peculiarly American trinity of the Sublime?

It could not; instead, the hopes of American science were blighted by the very nationalism and utilitarianism that was so loudly hailed. By the end of the period, according to a statement in the outline, the scientists were forced to realize:

> that there does exist in the country a deep, angry, sullen hatred of the concept of intellect maintained by the advocates of pure, unproductive science. The democracy and the religious community both sense that it is their enemy.

Among American historians, only Henry Adams used irony as successfully as Miller at his best. Like Miller, he controlled it better in his earlier work. In Adams' *History*, Jefferson's failures are generously allowed to illuminate Jefferson's greatness. In the *Education* and still more in the *Letter to American Teachers of History* the universal ironies illuminate Adams more than the universe. In Charles Beard and his followers, irony is present but often involuntary. One can never tell, and one suspects Beard did not know, what attitude he really took toward the mighty, ugly success story of industrial capitalism that dominates the *Rise of American Civilization*. There are a few glimpses in Turner of the ironies of the expanding frontier, and these are illuminated by Henry Nash Smith. For historians of the South and slavery, especially Vann Woodward (another fellow traveler with Niebuhr), irony is inescapable. But in most of the leading historians of the generation a decade younger than Miller, we see very little of it. Daniel Boorstin insists, in part convincingly, that America owes her successes to her avoidance of theory and her talent for improvisation. But he leaves out too much of the price of improvisation to convince readers who live in jerry-built cities

and worry about hand-to-mouth foreign policy. Richard Hofstadter, who is the most influential and perhaps the most brilliant of his generation of historians, falls short of irony for opposite reasons. His account of anti-intellectualism in American
religion and education slights the generous egalitarianism that in
both fields sometimes produced grotesque results.

Sometimes in this last book Miller's ironic paradoxes are forced
and his writing too high-pitched; the ironies are not always under control. Yet the book is full of suggestions for historians, and
full of the kind of insights that come only from deep knowledge
and long thought. Furthermore, and partly because of its incom-
pleteness, it helps us to approach the most interesting and complex of recent American historians. In this way, among others, it
illuminates the life of the mind in America. In his 1961 essay
"The Responsibility of Mind" Miller himself asked the most relevant question: "Is success the only goal?"

. 2 .

A Meditation on an
Unfashionable Book

I was one of many American historians who were deeply impressed by Reinhold Niebuhr's *Irony of American History* when it was published in 1952. For many years I assigned it to students, and therefore reread it yearly. I also looked closely, with students, at Niebuhr's other major works. In each college generation, some students admired Niebuhr and others detested him.

In the late 1960s in Berkeley campus conflicts became more and more bitter and violent, partly because of the immensely strong feelings of many students about the war in Vietnam, partly for other reasons including racial tension. The student upheaval of 1964, treated in essay 5, seemed in retrospect almost idyllic.

Reinhold Niebuhr was by this time out of favor. In the early sixties he had been seen as gloomy and guilt-ridden; in the late sixties he was attacked as an ideologist of the Cold War. This last charge was, I think, only partly true. Before his death in 1971, and before this essay was written, Niebuhr had become a stern critic of the war in Vietnam.

This article was published in *Christianity and Crisis,*

which had been founded by Niebuhr. I have restored certain cuts and changes made by the editors, but have otherwise left it unchanged. It seems to me especially important in this particular case to leave the article as it was written, so as to show how difficult I found it to apply Niebuhr's insights to this subject in that environment. Naturally the article received varying reactions, some of them sharply critical. It was published in *Christianity and Crisis,* XXVIII, 9 (May 27, 1968), pp. 120-22, and is reprinted by permission of the editors.

In 1952 many historians of the United States were deeply impressed by Reinhold Niebuhr's *The Irony of American History.* This was because the view it offered seemed to fit our circumstances. We had fought a war against evil forces, yet had come out of it covered with the guilt of Hiroshima. Despite official peace, there was no exit into the paradise of consumer's goods promised by wartime propaganda. Victorious, we confronted but could not overcome armed Communism, apparently spreading fast in Europe and Asia. Niebuhr warned us against surrender to either of two false programs: one to withdraw from the consequences of power into "innocence"; the other to make a final effort for "victory" and risk the nuclear destruction of humanity. It was part of the human situation, Niebuhr made us understand, that either innocent escape or durable victory was impossible. All triumph was illusory because of the dual nature of man. Good and evil, virtues and faults, were linked together. This was the ironical situation, true of all men, and especially poignant for Americans because of our tendency to insist on our own innocence. It was time, Niebuhr forcefully argued, to learn that we must act with full consciousness of the limitations of our powers and the necessary imperfection of the result.

For some, the underlying Christian neo-orthodoxy, for others the political wisdom in this view seemed to answer the needs of the time. American intellectuals, often brought up on the post-

Wilsonian disillusion, vigorously antifascist but dubious about the Four Freedoms and repelled by the horrors and lies of Stalinism, appalled in turn by the barbarous simplicities of McCarthyism, had plenty of reason to turn toward a doctrine of irony, ambiguity, and human limitations.

Today Niebuhr's teaching seems almost forgotten. The young clergy prefer the extraordinarily optimistic Harvey Cox version of Christianity as liberty indefinitely unfolding; hippies blame inhibition or bourgeois hypocrisy for all our troubles and urge us to drop out and turn on: in more traditional language, to be born again, once and for all, here and now. The far left demands that we learn to hate ourselves. A society that has degraded the Negro, that burns women and children in Viet Nam, has nothing worth saving: what it needs is destruction. Less obviously extreme, some historians are once more rewriting the past: just as revisionists once showed us we were wrong in the First World War and the Second, now revisionists show us that the Cold War was all our fault, need never have happened, and has led to the present disastrous situation.

In recent times, American culture has always sharply repudiated the ideas most recently in vogue. It is as hard for young people today to see anything good in the complacent 1950's as it was for young people in the 1930's to appreciate the self-indulgent, individualist 1920's. And indeed, when one looks at Niebuhr's volume today one can see flaws. There is not nearly enough recognition in some of his statements of the depth and variety of evil and failure in which Americans were involved at the beginning of the smug Eisenhower period. Statements about prosperity seemed to ignore the large exceptions; above all, "Americans" was used to mean "white Americans." Yet I find much in the book as alive and penetrating as ever; indeed, it contains a message that we badly need. If we don't like getting this message from the 1950's, after all, we can get it where Niebuhr got it, from the Old and New Testaments.

To start with, it is worth realizing that the guilt many Americans seem to be feeling for the first time is absolutely real and by no means new. To say, as historians have, that America lost her innocence in 1945 or 1917 or even 1861 is only a statement about how people felt: these are all dates at which certain groups woke up to ugly facts which had always been present. American society had been involved from the beginning in the extermination of one race and the enslavement of another. Callousness in warfare did not start in Viet Nam; we burned people alive more ruthlessly still in Korea in 1950 and above all in Japan in 1944-45, and very few liberals protested. In the complacent nineteenth century American troops had destroyed Indian villages. The ironies to which Niebuhr pointed could be found in the centers of the optimistic national legend: Puritans in search of spiritual freedom had scourged and hung Quakers; men who had gone to the frontier in search of individualism had persecuted many kinds of dissenters; believers in egalitarianism had condoned segregation and even lynching; Woodrow Wilson's New Freedom had tightened segregation in Washington; Franklin Roosevelt's war against Hitlerite racism had put citizens of Japanese origin in concentration camps. Worse, all these things had been done by people who believed in the purity of their own motives. There is nothing essentially new in the bureaucrats who really believe we are eventually going to bring the American Way of Life to South Viet Nam, or even in the generals who say with doubtless quite genuine regret that we must destroy a town in order to save it.

All this is true, and if our historians took more account of ambiguity they would be both more truthful and more interesting than they are. These sins and crimes are not minor blemishes on a virtuous culture, they are deeply part of it. Our crimes are almost always associated with our assumptions of innocence and virtue. Our New Israel, our New Freedom, our New Deal, our New Frontier always justify our crusades against our enemies, our Hiroshimas and Viet Nams. The future will make it all right: keep your perspective; look at the big picture.

Fortunately, as Niebuhr pointed out so well, we have never entirely believed our own rhetoric. Complacency is American; so fortunately is self-criticism. There have never been enough people to look into the dark corners, but there have always been many. Our saving remnants have been of many kinds. For the less terrible ills we have had thousands of organizers and committees: brave, energetic, often myopic, sometimes fanatical —above all indispensable. For most of these it was never necessary to abandon an essentially optimistic outlook; for the reformers their particular opponents: patent medicines, graft, bad housing, bad schools, the unearned increment, the trusts were artificial barriers to natural progress; enough organization and publicity could defeat them.

For the deeper evils, those more difficult to face, other kinds of prophets have been necessary; sometimes preachers like Weld or Rauschenbusch or Niebuhr; sometimes direct actionists, fanatical like John Jay Chapman or even insane and bloody like John Brown; sometimes practitioners of withdrawal like Thoreau or Randolph Bourne; often creative artists like Mark Twain or William Faulkner; sometimes odd people with elements of all these callings like Harriet Beecher Stowe.

What these dreadfully imperfect people, at their best, have had to offer is a combination of love and realism. Guilt after all is not enough. In religious terms, self-hatred must lead to acceptance and repentance. In social terms, the more we see what is bad in our society, the more we need to cherish all that is good. Liberals are at fault when they dismiss our faults in the time-honored manner as minor bad marks on a generally good report card. Radicals are at fault when they want to throw away as worthless all progressive traditions of an imperfect society. They are, in fact, wildly optimistic when they imply that self-hatred and destruction will somehow, in themselves, make things better; or when they find—as some of them, incredibly, still seem to do— innocence on the side of our opponents, purity among the guerrillas.

What we need to learn from Niebuhr's vision of American history, I think, is not only to see what is evil in the American past, but to see what is good, and how the opposites are linked. What practical difference will such a vision make? Progressives and liberals argue that appreciation of complexity will turn us all into Hamlets; we will become unable to choose and act at all. This is, indeed a danger, and perhaps there are omens of a cult of inaction in the hippie movement and in neo-isolationist proposals that America withdraw from the world. Similarly, optimists argue that a sense of sin leads people to sigh and weep and do nothing.

Fortunately, in the past neither complexity nor insight into evil has usually produced paralysis. It is interesting that many of those who earliest condemned and fought slavery were believers in inescapable human depravity. As Niebuhr points out, the President who presided with hard-pressed firmness over our greatest struggle had an unparalleled insight into the vanity of human wishes. And nobody has belonged to more committees, or wrestled with more immediate problems, than Niebuhr himself.

The more we can accept our own guilt and allow for that of everybody else the less, I believe, we are likely to move from one mistake into its opposite. Concretely, we must immediately stop slaughtering civilians in Viet Nam; here the young prophets have been more right than the middle-aged sceptics. But in stopping, we must not be callous or blind to the danger that other civilians will be slaughtered by our victorious enemies. We must admit our final responsibility for the Negro ghettoes and understand the anger of those who live in them, but we must not delude ourselves that things will be better with more fires or more race war. It is by now impossible to shut our eyes, and it is no good to beat our breasts. Once and for all—and this is what Niebuhr told us in 1952—we must live with ourselves as we are. If we are to get away from the cycle of crusade and frustration we must cherish, even in the midst of guilt, all the creative resources of our culture.

· 3 ·

Philip Greven and the History of Temperament

In the 1970s American historians were drawn as never
before into arguments about historical theory and
method. This was largely because of the impact of new
kinds of history, including psychohistory and the "new
social history" coming from France. This review essay is
an attempt at sympathetic criticism of a book which has
elements of both the new varieties of history mentioned
above, and also of religious history.

Philip Greven's *The Protestant Temperament: Patterns of Child-Rearing, Religious Experience, and the
Self in Early America* was published by Knopf in 1977.
This review article was published in the *American
Quarterly*, XXI, 1 (Spring 1979), pp. 109-15. It is reprinted by permission of the editors and is copyright,
1979, by the trustees of the University of Pennsylvania.

As most interested readers know by now, Professor Greven's
recent book is an ambitious effort to explain the nature of early
American culture in terms of three temperaments, the "evangelical," the "moderate," and the "genteel." Arguing eloquently

that we must study the history of the inner self in order to understand behavior, Greven defines temperaments as "patterns of feelings, thought, and sensibility." These all-important patterns are formed in childhood; therefore historians of temperament must concentrate their research on childhood experience.

Far the largest section of the book—and one which constitutes an important historical contribution in itself—centers on the childhood experience and consequent temperamental patterns of the people Greven calls evangelicals. Evangelical parents approached the solemn duty of raising their children in a mood compounded of worried love and agonizing fear for the consequences of inherited evil. The self and especially the body, its mortal habitat, were to be suspected and suppressed. The total authority of God in his world was to be reflected in the total authority of human parents. So that the children might have some chance of eternal happiness, no pains must be spared in the breaking of their wills.

Children brought up in this style learned rigorously to suppress their feelings of rebelliousness and anger against God and their parents. By this suppression they accumulated formidable reserves of hostility which were eventually redirected against those who differed with them in adult life. Rigidity and intolerance toward oneself left no possibility of flexibility toward others.

Moving beyond this fairly familiar territory in a daring direction, Greven argues with a good deal of telling quotation that evangelical boys were *feminized*. They were required rigidly to repress not only male sexuality, but also all the robust and aggressive tendencies traditionally attributed to males. They were forced to become—the metaphor was often used for males—brides of Christ: meek, submissive, and pure. Only when this drastic goal was achieved, when the self was totally subdued, could evangelicals begin to feel some hope for their own salvation and even for the conversion and transformation of the world. The picture is hardly a friendly one, yet it is far more than caricature. Many of us will find elements in it that are familiar enough, from books about the past and even from scars we bear ourselves and recognize on those around us.

Greven's second category, the moderates, are like most intermediate categories harder to describe in bold dramatic terms. Devoted to the ideal of balance in all things, they believed that even God governed rationally and according to rules. In imitation of this pattern, moderates were authoritative rather than authoritarian parents. The child's will was to be gradually bent but not broken; he was to be motivated by love as well as duty. Religious instruction emphasized growth in grace rather than the necessity of despair followed by violent conversion. Sexuality, like the passions of ambition, avarice, and anger, could be channelled and bounded rather than totally suppressed. Moderate males were allowed to mix some feminine sensibilities with their clearly dominant masculinity.

Moderate families tended to be both extensive and rooted in a locality; parental absolutism was mitigated by the presence of servants and, still more, grandparents, whose indulgent attitudes were deeply feared by evangelicals. Since they believed that sin and virtue were necessarily mixed together in all human beings, moderates welcomed variety. Society should be organic and hierarchical, the church comprehensive rather than separated or purified.

For the genteel, Greven's final category, family pride was the central emotional commitment. In their mansions the actual hard work of child-raising was left to nurses and other servants, leaving parents free to be at once indulgent and a little aloof. Small boys were dressed and treated just like their sisters until about the age of six, when masculinity was expected to be assumed along with trousers. From then on pride and indomitable will were to be neither destroyed nor bounded, but accepted as proper signs of high spirits. Among genteel adults, tenderness and aesthetic sensibility were entirely acceptable counterparts of taken-for-granted masculine courage. For women, a due sub-

ordination was not to mean a lack of elegance. Genteel religion
tended to be formal and public; fervent piety and inner enquiry
were disapproved. The self, like the organization of society, was
to be accepted without much question.

Mr. Greven finds these divisions to be useful for the seventeenth, eighteenth, and early nineteenth centuries, but after his
general description he throws in a special discussion of the
American Revolution as an illustration of the importance of
temperament. The Revolutionary earthquake, like other social
upheavals, erupted along fault lines which had long been evident.
The choices people made "involved the entire self, and thus
became a matter of temperament." Evangelicals, who knew tyranny first-hand and were full of unexpressed resentment, were
fiercely republican. Given to suspicions of conspiracy, they
feared both grasping authority and the seductions of effeminate
luxury. Moderates had a harder choice to make, since liberty for
them consisted of a due balance between order and freedom.
Many of them, however, concluded that the balance was tipping
too far in the direction of authority and therefore, with or without misgivings, supported the Revolution. For the genteel,
authority was necessary and its distribution about as it had to
be; many of them in crisis supported the crown.

Mr. Greven's book is deeply if unevenly researched; his prose
is relaxed and fluent; above all, the book is full of fresh and
fascinating illustrations, some of which stay in one's mind long
after one has put aside the book. Because it is daring, Greven's
argument has its flaws. To point to some of these is not to deny
the book's power and importance.

In the first place, an examination of Greven's footnotes shows
that his sampling of sources is not really adequate to support
conclusions about American culture over two and a half centuries. In an honest and disarming note (page 404) he describes
how his inquiry evolved, beginning with the evangelicals and
looking far less hard at the genteel. Like many writers of colonial

intellectual and religious history he concentrates very heavily on the supremely articulate Congregational clergy of eighteenth-century New England, and to a lesser extent on their Presbyterian brethren of the middle colonies. In addition to these, his evangelicals include a good sprinkling of seventeenth-century Puritans and Quakers. Wesley and Whitefield, neither American, play a prominent role. So does one lone nineteenth-century Baptist intellectual, Francis Wayland.

Among moderates also, eighteenth-century Congregationalists and Presbyterians are in an overwhelming majority. This category also includes covenant-minded Puritans of the seventeenth century (here Greven follows Perry Miller rather than his critics), some eighteenth-century Quakers and Episcopalians, and a few of the emerging Unitarians of the beginning of the nineteenth century. The genteel are divided about evenly between Congregationalists and Episcopalians, with the addition of a few liberal Quakers. Except for William Byrd, all are people of the late eighteenth century.

Thus chronology, neglected by Mr. Greven, forces its way into the minds of his readers. The evangelical temperament, formed in the sevententh century, was strong at least through the nineteenth.[1] It is possible that moderates were especially important in the eighteenth century, with forerunners in the seventeenth. It seems likely that the genteel flourished mostly in the period shortly before the Revolution, when as many historians have pointed out British modes were gaining strength among the top colonial class. Mr. Greven's analysis cannot possibly carry him convincingly past the second great religious upheaval in America, the Revival of 1800. One cannot deal with even the very beginning of the nineteenth century without Kentuckians and Tennesseeans, without the dissident Baptists and rebellious Presbyterians, above all without American Methodists, who were hardly much like their English founder.

Geographical differences, to a lesser extent, also demand more consideration than they are given. For instance, Mr. Greven is not quite consistent as to whether Southern gentility is the same as Northern. A closer look at the rich manuscript materials available for Virginia would have helped here. It would, I think, have enabled Greven to handle convincingly the surface discrepancy of aristocratic support for the Revolution.

The book is primarily about eighteenth-century society and religion in the Northern and Middle Atlantic colonies and states. That is a big topic, and I do not mean to fault Mr. Greven for not having worked hard enough. He has worked very hard, but in places his claims outrun his researches.

Partly because of the book's actual chronological and geographical emphases, the names of his categories do not always convey his meanings. By the term "evangelicals" Mr. Greven, like Alan Heimert whom he admires with some qualifications, means fervent and millenarian Calvinists. I am not sure that seventeenth-century Quakers or late eighteenth-century Arminian evangelicals fit his analysis. Were there not at least degrees in the self-hatred taught to children of these groups?

In dealing with the Calvinists themselves, Mr. Greven perhaps emphasizes repression a bit too exclusively. Though he occasionally recognizes the strength given the personality by the destruction of self, his examples seem to me not sufficiently to convey the feelings of joy, peace, and love which, in so many accounts, succeed the deadly struggle with the will. A more balanced account of these not very balanced people might have helped to illuminate still more clearly the seeming paradox of children whose wills were broken growing up with enough confidence to exert authority in turn over their own children in the same total manner. If this had not happened, the evangelical temperament as Greven describes it could have lasted only one generation. The Calvinist revolutionary, the authoritarian rebel, is really neither paradoxical nor unfamiliar, but in dealing with him Mr. Greven might have looked still more deeply within.

Mr. Greven knows that it is hard to set firm boundaries to the category "moderate." Ranging from John Winthrop to Thomas

Jefferson, his moderates tend naturally enough to drift one way or the other. Greven uses the word "genteel" in a surprising way. At least since Santayana's great essay on the Genteel Tradition the word has tended to connote pallid middle-class respectability and not, as in Greven's book, high-handed confident aristocracy. As all these cavils suggest, Mr. Greven's categories are not parallel: "evangelical" is a religious category, "genteel" as he uses it is a class term. Only "moderate" remains really descriptive of temperament alone.

Like most recent historians, Mr. Greven fails to get very far with causation. Many will find it all to his credit that he grinds no particularly theoretical ax. He is not, to start with, a religious historian in the sense of one who regards religious categories or allegiances as always or even usually determinative. Only in his evangelical category is religious experience central in people's lives. Among moderates religion is moderately important; for the genteel its role is clearly minor.

Again, Mr. Greven is not really a childhood or family historian. In associating child-raising practices with religious commitment, he raises—possibly on purpose—a chicken-and-egg problem. At least once he says clearly that belief in Original Sin and innate depravity were grounded in the feelings of evangelicals toward their inner selves (and thus in their early upbringing). Yet it seems at least equally clear that the early upbringing of children was affected by the religious allegiance of their parents. This in turn was sometimes changed partly for secular reasons: we know that prosperous Quakers and Presbyterians often switched to the Church of England. Did such converts change their inner feelings and therefore the way they raised their children? At times Mr. Greven's division between evangelicals and moderates corresponds very closely to the familiar Troeltschian distinction between church and sect; surely whether one was a come-outer or a formalist had something to do with inherited tradition as well as inner makeup.

Still less is Mr. Greven a historian of ideas. Religious ideas get little close analysis, and secular ideas are almost entirely neglected. No doubt religion was extremely powerful in early America, but not everybody thought in religious terms all the time. Fully to explain the moderates' love of balance, one needs to draw on the English Enlightenment and its Augustan, Renaissance, and classical background. Among revolutionaries, one wonders how Mr. Greven would classify Thomas Paine or, if he is ruled out as British, his most radical American followers from 1776 to 1795. Surely these firebrands were neither moderate nor genteel, yet they violently opposed Calvinist evangelicals. Was there a deist mode of bringing up children? It is a little awkward when Mr. Greven has the candor to point out that Locke, the grand intellectual patron of moderation, was "almost as repressive" in his child-raising precepts as Susanna Wesley.

Mr. Greven is often at his best when he is being boldest in his psychological interpretations. Yet, though he gives great emphasis to the effects of childhood experience, he is hardly a psychoanalytic historian. Training forms the psyche, and training is clearly affected by resources, class position, and tradition. The moderates have stability and servants; the genteel, rich traditions of family pride. As in the case of changing religious allegiances, one wonders whether child-raising practices changed when people moved upward in the relatively mobile colonial society. Mr. Greven seems perhaps least of all an economic determinist, and yet there are passages in his book that make one wonder whether much would be changed if the three categories were named lower middle class, upper middle class, and aristocratic.[2]

Clearly much more thinking needs to be done before Mr. Greven's categories are accepted as conclusive. Having said that, I want to add that Mr. Greven's openness, flexibility, and tentativeness are the main strength and not the weakness of the book. An admirer of William James, Mr. Greven knows that categories are tentative, and must be tested and revised as they are used. His refusal to be dogmatic about causation will alienate evangelical Freudians and Marxists, but may well interest the more catholic members of both these camps. My only criticism of his tentative-ness and the fluidity of his categories is that he might have admit-ted them more explicitly. This would have made it easier for others to go on from where he leaves off.

Putting aside for the moment the strengths and weaknesses of this important book, I should like to raise the question of the place of history of temperament in historiographical develop-ment. Let us define it as the study of groups differentiated by psychological traits, with the problem of causation left unsolved. Historians of temperament deal with inchoate emotions pri-vately expressed or demonstrated in action more than with articulate and organized ideas, though they do not necessarily neglect the latter. In America at least, they necessarily give much of their attention to religious feeling and its expression. They take account of social status, but do not necessarily begin with it.

Looking for analogies and parallels, one might start with the influential and often brilliant work of French historians of *mentalité*. There are however two major differences between their work and that of Mr. Greven. First, historians of *mentalité* are much more cautious than Greven about both geographical and chronological divisions. Second, the French historians and their close followers start as a matter of course with "structures," geographical, demographic, and economic. Sometimes even the best of them find it difficult to move clearly from this starting point to consciousness, expression, and events.

It is interesting that in his earlier major work,[3] Mr. Greven constantly cites British and French demographic and other his-torians, while in the present book European writers are hardly mentioned. In the earlier book, Mr. Greven starts with solid demographic data, and in his present book he seems to give at least equal causal status to psychological and religious categories. It may well be that just at present this is the best tactic, and one for which American historians, with their pragmatic and eclectic tradition, may be especially fitted. Much of the social science, particularly anthropology, that is at present most influential among American historians suggests a very open and pluralist approach to the relation between culture and socio-economic fact.[4]

History of temperament probably has a future in America; it certainly has a past, though it has not been clearly perceived as a historical form. Among recent American historians, Greven is particularly influenced by Alan Heimert. As Greven points out, Heimert's *Religion and the American Mind* received very mixed reviews, partly because of certain obvious exaggerations. Yet Heimert, dealing with the period from the Great Awakening to the Revolution, called attention to two important religious-psychological types: the fiercely egalitarian and revolutionary Calvinist and the cool, polished, essentially moderate Arminian. These types do not include every individual in eighteenth-century America, and the political stance of each was less predictable than Heimert suggested. Moreover Heimert failed to extend the dichotomy convincingly to the end of the century. Yet both types really existed, and Heimert found plenty of instances of each. Heimert's exposition of both is now supported and deepened by Greven.

Another historian who comes to mind, though Greven does not refer to him, is Richard Hofstadter. Like Greven, Hofstadter was hard to pin down on causation; like Greven, he was deeply concerned with the historical mixture of psyche, religion, idea, and class. Some of his students followed him with greater or equal success; some with less. Marvin Meyers' "Jacksonian persuasion" seems a kind of category fairly similar to Greven's temperaments. One could go on with many other historians of many kinds and origins who have been willing—for some tastes too willing—to discuss the Mind of the South, or the ways of People of Plenty, without getting into sharp debate about causation. Such efforts at history of temperament have drawn solid criticism, but few would want to be without the best of them.

Historians will certainly continue to discuss Greven as they have his predecessors in this field. One question that will be asked is whether, as he sometimes hints, his three categories can be pushed forward into later periods. A few guesses may be allowable. As I have already suggested, Greven's third psycho-social type, the confident, passionate aristocrat, has been recessive since the Revolution. In tidewater Virginia it survived as an ideal; perhaps some of its contours are recognizable in the New York that produced Franklin Roosevelt. His other two categories, evangelicals and moderates, seem far more durable. Let us think, for instance, of late nineteenth-century reformers in something like his terms. One type was egalitarian, populistic, combative, apocalyptic: very possibly these might prove to have evangelical origins and upbringings. Another was moderate, reformist, polite in language, gradualist, optimistic: such people may well have been formed by liberal religion. Similar categories spring to mind for still more recent periods. Obviously, continuities will be only partial; for new periods research is needed at least as broad and deep as Greven's own.

Historians who want to deal successfully with temperament need three qualifications. First, they must be bold; second, they must be diligent in research; third, they must be flexible—their categories must not be too easily reified; they must be willing to recognize groupings they did not expect to find. Greven seems to me to exemplify these three traits very well. He is almost as bold as Hofstadter and harder-working; he is more flexible than Heimert. One might be tempted to add a fourth qualification: caution. Here Greven would not fare as well. Yet caution is never the first necessity; mistakes can be corrected and something will still remain from an enterprise like Greven's.

The principal test of this kind of book can only be the pragmatic one: do Greven's categories, regardless of their causal clarity or precision of boundaries, help us to understand better the events of his period? I think they do, especially the "evangelical" grouping. Neither strictly religious nor strictly social-economic explanations will suffice to make clear why certain kinds of American religious spokesmen were, within the space of thirty years, equally violent against bishops, Tories, Jacobins, and Jeffersonian Republicans. A general temper, a psychic bent, a rhetorical habit may well take us further—at least for the present.

In the long run, historians will doubtless try, as they always have, to push further their search for solid, impeccable causes. Meanwhile, history of temperament can serve as a fine staging area for the gathering of forces. Greven's book is an important and above all an interesting one. As this review may not have made clear, its interest lies as much in its examples as in its generalizations. The book's vitality comes from the fact that it deals at all points with real human beings, and not with mere clusters of ideas and interests.

Notes

1. Lawrence Stone, in his monumental *The Family, Sex and Marriage in England, 1500-1800* (N.Y.: Harper and Row, 1977), finds the period from 1670 to about 1790, at least among the upper-middle and upper classes, an interlude of permissiveness between two ages of authority and repression. Chronological comparisons of American development would say a lot about differences between two related societies.
2. In a footnote on page 12, Mr. Greven raises just this question, only to dismiss it.
3. *Four Generations: Population, Land, and Family in Colonial Andover, Massachusetts* (Ithaca: Cornell Univ. Press, 1970).
4. In his 1978 presidential address to the American Historical Association W. J. Bouwsma discusses this approach in strongly affirmative terms, and refers at length to its anthropological sources. Bouwsma, "The Renaissance and the Drama of Western History," *American Historical Review*, LXXXIV (1979), 1-15.

Index